ESSENTIALS OF COMPUTATIONAL FLUID DYNAMICS

ESSENTIALS OF COMPUTATIONAL FLUID DYNAMICS

Jens-Dominik Müller

CRC Press is an imprint of the
Taylor & Francis Group, an **informa** business

Lotus Formula 1 race car, with streamlines coloured by pressure coefficient.

Simulation performed with STAR-CCM+ of CD-Adapco.

(Image courtesy of Dr. N. Forsythe, Lotus F1).

CRC Press
Taylor & Francis Group
6000 Broken Sound Parkway NW, Suite 300
Boca Raton, FL 33487-2742

Printed on acid-free paper
Version Date: 20151013

International Standard Book Number-13: 978-1-4822-2730-7 (Paperback)

Visit the Taylor & Francis Web site at
http://www.taylorandfrancis.com

and the CRC Press Web site at
http://www.crcpress.com

Contents

Foreword

This book is the evolution of extended lecture notes that have grown out of my teaching of CFD to undergraduate students. All of the relevant texts that I considered approached the material from the perspective of a CFD code developer, a role that the early users of CFD had to adopt. Today's CFD codes are very robust and user-friendly, to an extent that many students only ever look at the tutorials provided by the code vendors or online videos by fellow practitioners.

This text is written for undergraduate teaching to provide the background that a user of a commercial or open-source CFD package would need to competently operate it and analyse the results. The level of material included — and what was omitted — has been shaped by years of student questions of CFD beginners. What this text focuses on is driven by students' responses to practical coursework, and the limits of how much material can fit into a first course on CFD.

I'd like to acknowledge particularly the help by my PhD students Siamak Abkarzadeh, Mateusz Gugala, Jan Hückelheim, Yang Wang and Xingchen Zhang who helped proofreading the manuscript and made many valuable suggestions. The SnappyHex mesh for the Ahmed body was produced by Mohamed Gowely.

1

Introduction

1.1 CFD, the virtual wind tunnel

Computational Fluid Dynamics is often called the virtual wind tunnel. And while it offers much more than just a "testing facility" on the computer, its main use is certainly in the analysis of flow around a given design and there are clear analogies. There are a number of advantages of the virtual wind tunnel CFD over the physical wind tunnel:

Cost: unless your company has enough flow experiments to run to afford and occupy their own wind tunnel, it is typically much less expensive to perform a CFD simulation as compared to producing a model and testing it in a tunnel.

Turnaround time: results can be obtained much more rapidly as no model has to be built. Industry often works on a turnaround time of 12h, overnight, for moderate to high levels of accuracy and detail.

Exact similarity: models normally have to be tested on reduced scale in an experiment. This means that in general the tunnel speed is fixed to match one similarity parameter, e.g., the Mach number, while the other relevant similarity parameters are not correct, leading to errors e.g., in the measurement of viscous effects if the Reynolds number is not kept constant. CFD simulations can be performed at the original scale; all similarity parameters can be matched.

Non-intrusive: measuring the flow in experiments is invariably intrusive: the probe or the suspension of the model perturbs the flow and results in measurement errors. CFD requires neither probes for measurement nor stings or wires to suspend the model.

Analysis: experimental testing requires careful planning to locate probes and record the data of interest; in general, data are only available in isolated points or in 2-D cross-sectional planes. In CFD the flow solution in the entire simulation area is available and open to flow visualisation and analysis.

Feasibility: not all flows can be measured experimentally, think, e.g., of reentry flow of the space shuttle. With a suitable model CFD can simulate any flow, e.g., reentry flows have been simulated using a model for the real gas effects of hypersonic flows.

Extendability: a wind tunnel is just a wind tunnel; it only analyses the flow over a given surface. CFD on the other hand can be embedded in a chain of tools, or in an automated design loop where the surface is changed by the design system to improve performance.

These advantages have led to a major increase in the use of CFD in design of surfaces in flow over the past 20 years. For example, in aeronautical design 20 years ago a large number of wind tunnel testing campaigns were conducted during the development cycle of an aircraft. Today wind tunnel testing is reduced to 2-5 campaigns for each component.

However, experimental testing is still indispensable, as CFD has some major drawbacks:

Turbulence: turbulence is an inherently 3-D and unsteady phenomenon that involves very small flow features at high Reynolds numbers. Today it is not possible to simulate all that detail. Even if the current growth of the hardware were to continue unabatedly as Moore's law suggests, it would take another millennium to have sufficient power to simulate fully turbulent flow over a passenger aircraft in cruise condition. Effects of turbulence have to be modelled for highly turbulent flows and the state of the art in turbulence modelling is rather limited.

Accuracy: besides turbulence there are other very small flow features which have importance for some cases and cannot be simulated in enough detail on today's hardware.

Error control: the errors which are incurred from insufficient detailing or from poor modelling currently cannot be quantified. That is, given a CFD solution there is no way of assuring that the errors in it are below a chosen threshold. It takes a lot of user experience to estimate the errors and to calibrate and correct the results. For some flows CFD will only give qualitative answers.

Notwithstanding the problems of errors, CFD is a key technology in design and process analysis in a large variety of areas. The following section will highlight some applications.

1.2 Examples of CFD applications

The earliest applications of CFD were in aeronautics, and aeronautical applications were the main drivers behind the progress in CFD. Had the Wright brothers had access to CFD, they might have designed the Flyer somewhat differently (Fig. 1.1), although their intuition served them well for their simple low-speed design.

FIGURE 1.1
Contours of pressure on the mid-plane of the Flyer. (Courtesy of Fluent.)

Today's aircraft are much more complex and the transonic flight regime is very challenging. An example is the simulation of flow over a wing with slat and flap deployed (Fig. 1.2). The lifting wing produces higher pressure below, lower pressure above the wing. At the tip a swirling motion results from the air being able to follow this pressure gradient. The accurate simulation of the wingtip vortices is important for a number of reasons. On the one hand the kinetic energy absorbed by the swirling motion of the vortices represents a significant amount of drag which should be minimised during takeoff. These vortices can persist a long time after the aircraft passed and then represent a danger to a smaller aircraft flying through them. The wing geometry of large aircraft such as the Airbus A380 is designed to induce a vortex pattern that contains instabilities which result in a faster breakdown. During landing the flaps will be extended to a maximum, which creates a lot of aero-acoustic perturbations that need to be minimised to reduce noise pollution.

Modern aircraft not only have to perform well aerodynamically, they also have to meet more and more stringent environmental regulations. The various aspects of the flow through an aircraft engine are studied extensively using CFD to maximise performance and minimise pollution. A particularly challenging aspect is the flow through the combustor where air pressurised by the turbine is mixed with fuel and ignited. The combustion of fuel and air is enhanced by the mixing action of turbulence, and the released heat of the combustion in turn enhances turbulence; both phenomena are very strongly coupled with each other. A flow simulation in the combustor is shown in Fig. 1.5. Fuel injectors are shown at the lower left edge of the annular plane,

FIGURE 1.2
Wingtip vortex on a wing with slat and flap. (Courtesy of NASA Ames.)

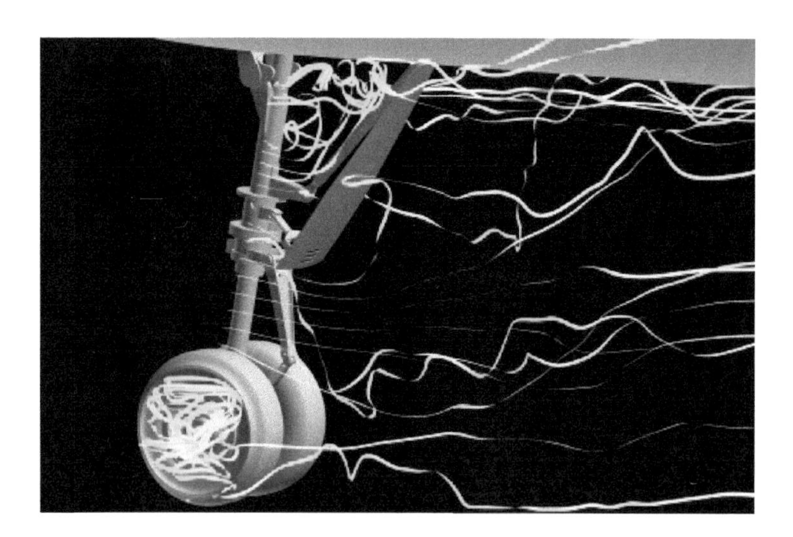

FIGURE 1.3
Landing gear, LES simulations of unsteady turbulence, pathlines. (Courtesy of NASA/The FUN3D development team.)

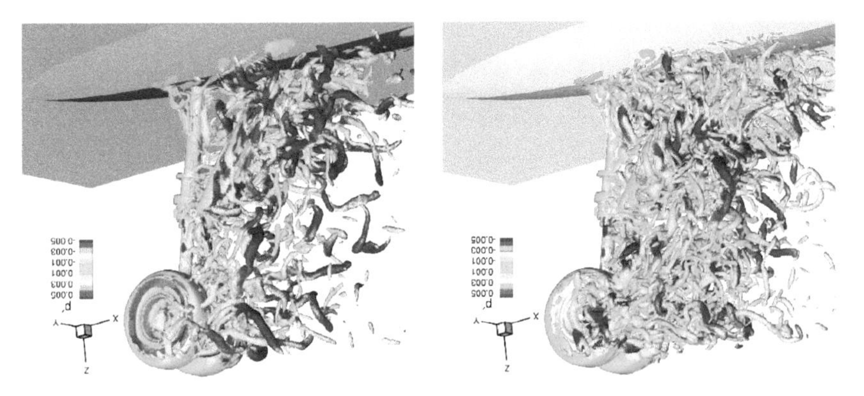

FIGURE 1.4
Landing gear. LES simulations of unsteady turbulence, visualisation of the turbulent vortex cores on a mesh with 9 million cells (left) and 25 million cells (right). (Courtesy of NASA/The FUN3D development team.)

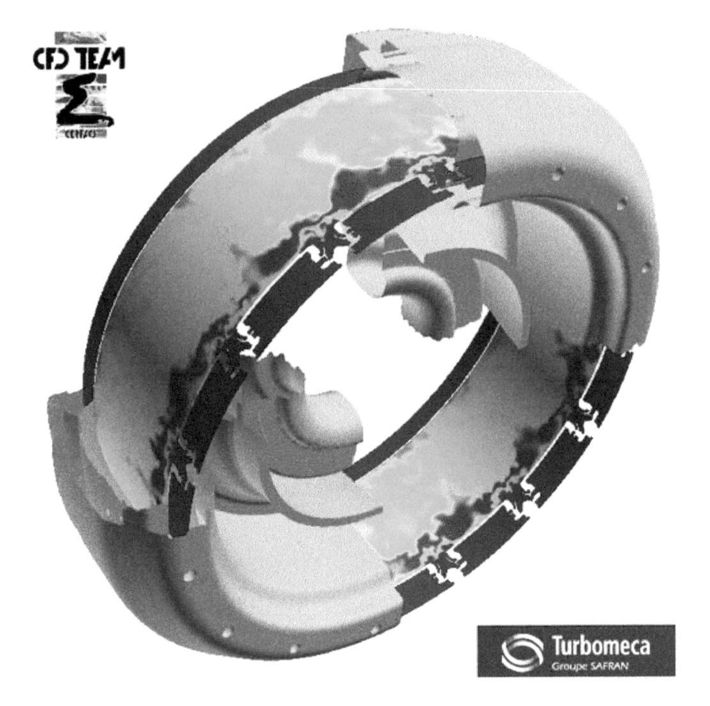

FIGURE 1.5
Large Eddy Simulation of turbulent combustion in an annular combustion chamber of a helicopter engine. (Courtesy of CERFACS.)

FIGURE 1.6
External flow around the body and inside the engine bay of the Bentley Arnage: contours of static pressure and path lines coloured by velocity magnitude. (Courtesy of Fluent.)

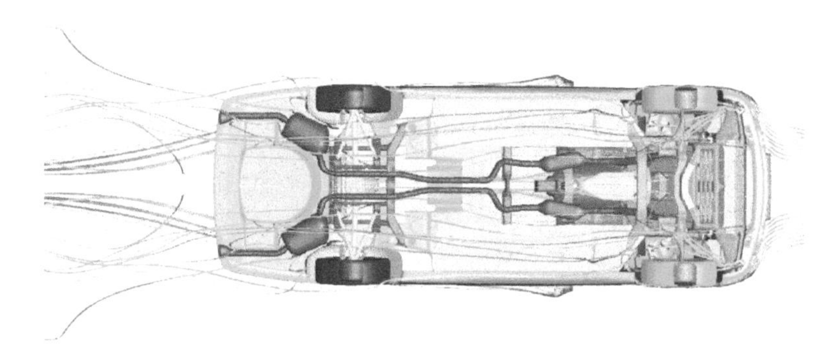

FIGURE 1.7
Path lines coloured by velocity magnitude showing the flow pattern near the underbody of the Bentley Arnage. (Courtesy of Fluent.)

mixed with air and strongly swirled to stabilise the flame. The combustion heats up the gas (red colours); the turbulent mixing of the hot gases with the remaining air can clearly be seen.

The turbulence of the flow is simulated using the Large Eddy Simulation approach (LES), where the largest vortical structures (*'eddies'*) that carry the majority of the turbulent kinetic energy are resolved in detail in the simulation, while smaller scales are averaged and modelled with heuristic models.

CFD is also applied extensively in the design of (usually) non-flying vehicles. Figures 1.6 and 1.7 show the flow simulation around a car. A non-negligible part of the drag comes from the flow between the ground and the underside of the car.

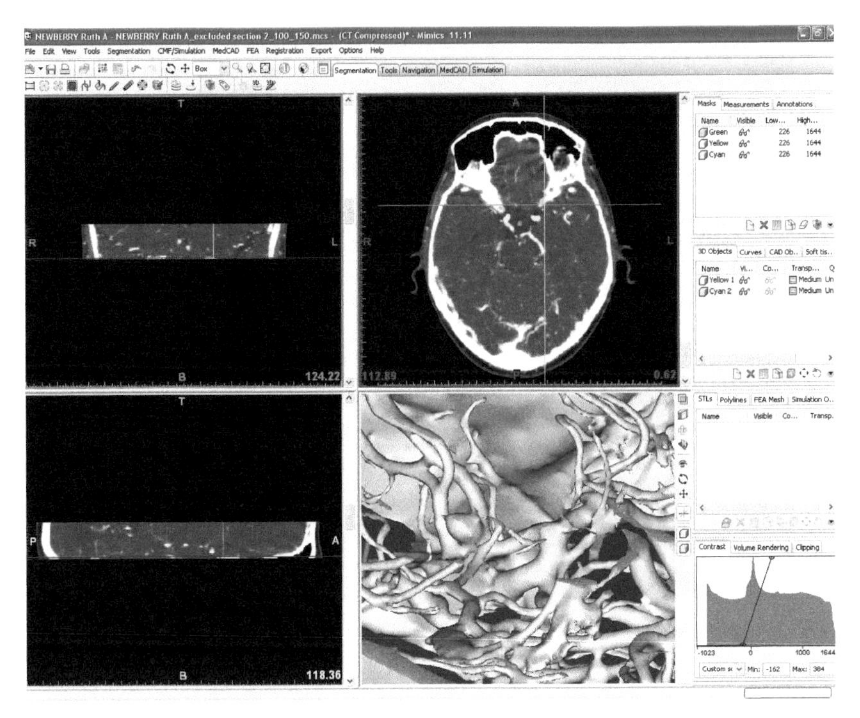

FIGURE 1.8
Image segmentation: CT scan image of a cerebral aneurysm.

CFD is an important tool to investigate biological flows as experiments are difficult or impossible to conduct. Figures 1.8 to 1.11 show an aneurysm on a cerebral blood vessel, a bulge in the vessel wall which can potentially rupture with fatal consequences. The geometry of the aneurysm is modelled from a patient's CT scan (Fig. 1.8). The CT image is then "segmented" to extract the geometry of the blood vessel that is to be simulated (Fig. 1.9), leading to a geometry model (Fig. 1.10) that can be fitted with a mesh for a CFD computation. In this case the inflow velocities toward the aneurysm location can be measured specifically for this patient using transcranial Doppler ultrasound techniques; however, the outflow velocities are not accessible to an accurate measurement and have to be assumed. The simulation shows that the flow pattern in the aneurysm is highly sensitive to the choice of outflow velocity.

CFD is also used in athletics. Fig. 1.12 shows the effect of replacing a spoked rear wheel with a disk rear wheel. Fig. 1.13 shows the flow around a swimmer's arm and hand. The analysis helps to understand the flow and to improve the athlete's technique.

Applications of CFD are most numerous. A good start is the website of the Fluent code, www.fluent.com. Another portal with excellent resources for CFD is www.cfd-online.com.

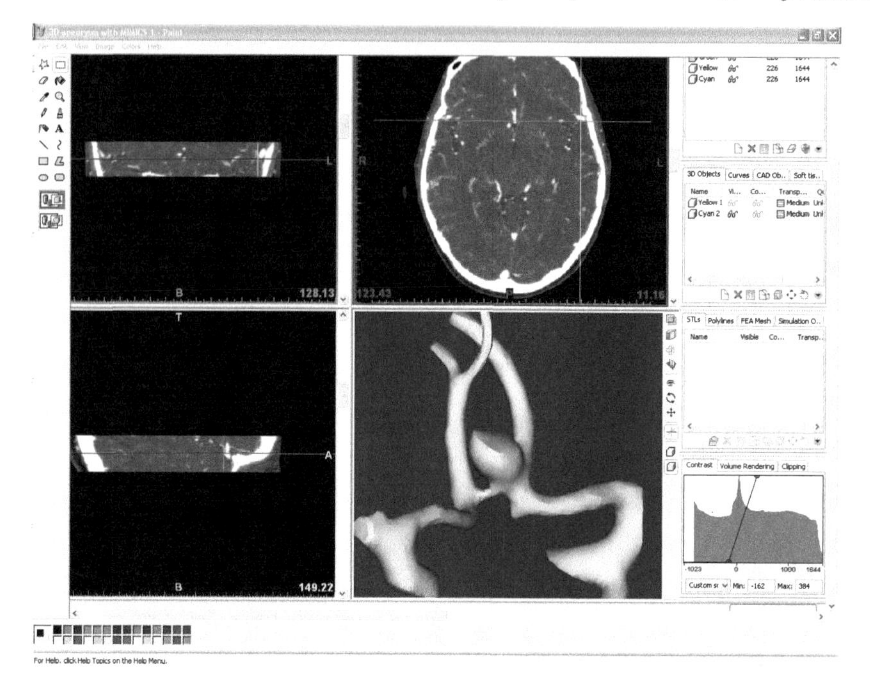

FIGURE 1.9
Image segmentation: the geometry of the blood vessels is extracted by selecting an appropriate level of grayscale in the CT scan image. (Mimics segmentation software provided by Materialise.)

1.3 Prerequisites

The state of the art of CFD has evolved significantly over the past 20 years. Then it was a black art mastered by highly trained specialists using quite frequently software written by themselves. To obtain an accurate solution one was required to have good knowledge in mathematics, numerical methods and flow physics. Today it has developed to be in most cases a routine engineering task accessible to engineers with, ideally, a good understanding of the flow physics. Operating a CFD package can be learnt in a 2-day course.

However, unlike e.g., finite element analysis in structural mechanics, there still is no way to rigorously quantify the errors of the CFD solution. In many cases the solutions are only qualitative, i.e., the flow phenomenon of the case is represented correctly but with large quantitative errors. For example, this is very often the case in unsteady combustion simulations. It is still up to the user's experience in CFD and his knowledge of the flow physics to judge the accuracy of the CFD prediction.

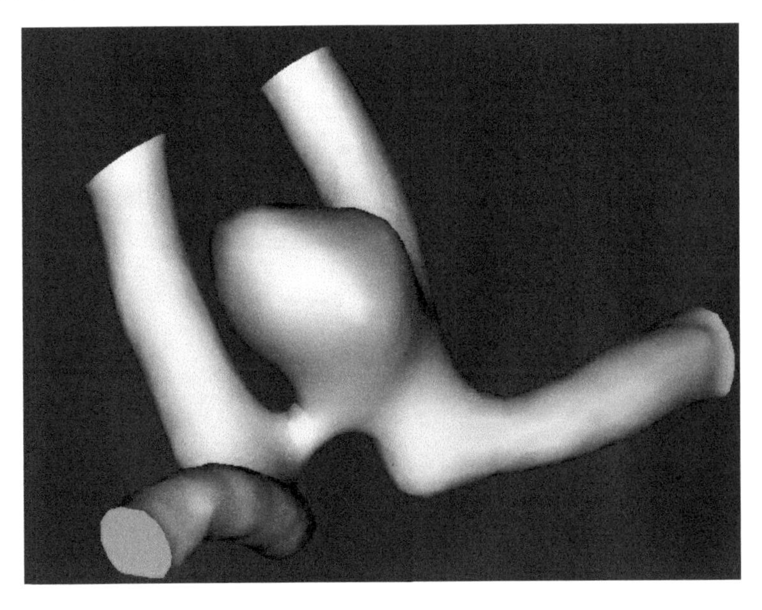

FIGURE 1.10
Image segmentation: geometry model.

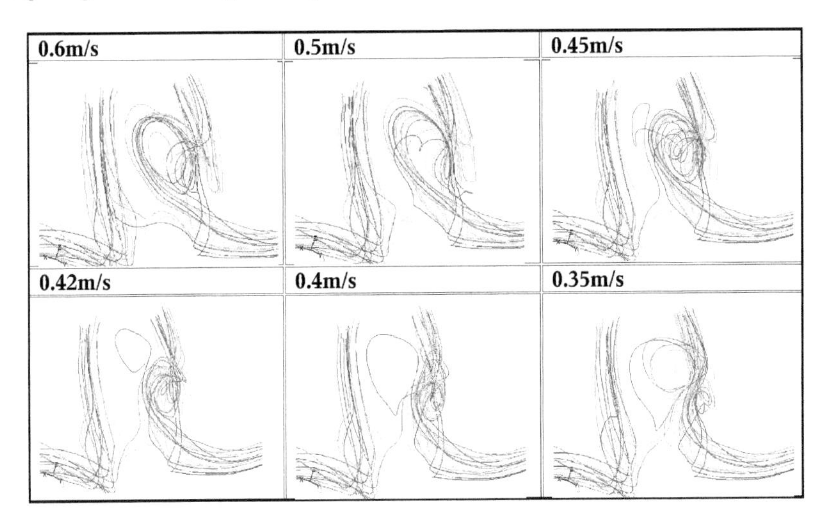

FIGURE 1.11
Effect of outflow velocity on flow patterns in a cerebral aneurysm.

This guide expects the reader to have a good knowledge of fluid dynamics. It will cover the material in a quite general fashion and concentrate on basic principles.

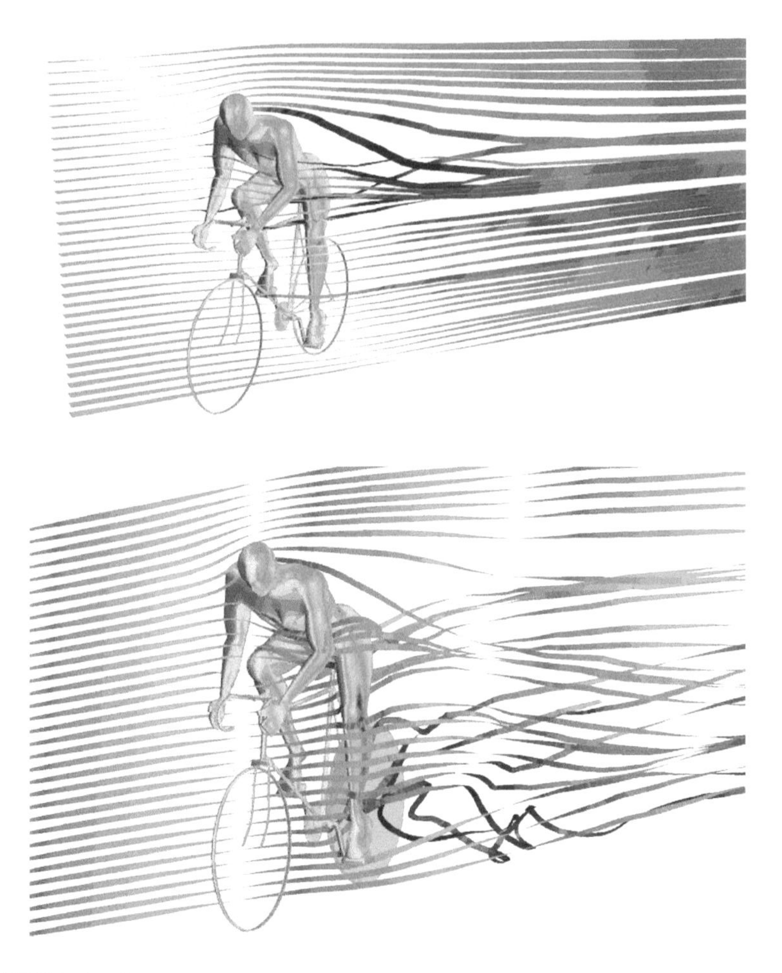

FIGURE 1.12
Flow path lines around a cyclist with a spoked rear wheel in a 20 mph cross-wind (top) and a disk rear wheel (bottom). (Courtesy of Fluent.)

Only a very small amount of the mathematical and numerical theory is considered in this guide, just as much as it is relevant to the operation of a modern commercial CFD package. The reader will learn to choose appropriate mathematical models and suitable numerical discretisations, to set boundary conditions, to choose an appropriate time-stepping scheme. It will briefly introduce the major mesh generation techniques and discuss mesh quality. The final section discusses how to analyse the results and assess the errors in the solution.

This guide will leave many questions open. Answers to those can be found in the references given in Section 1.4. Hopefully, however, all the aspects nec-

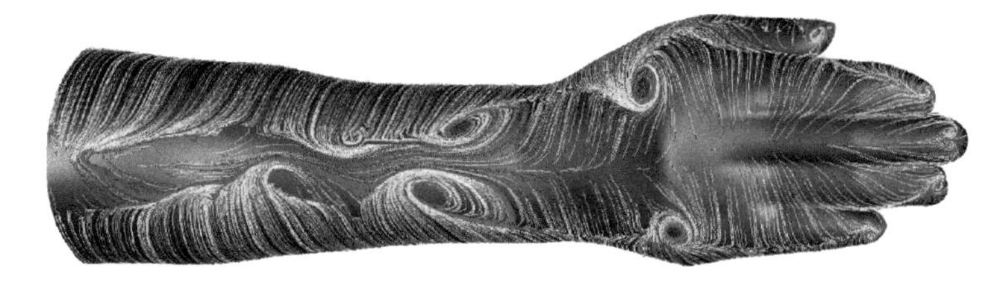

FIGURE 1.13
Position of the hand/arm at the beginning of the propulsive phase of the freestyle stroke, and oilfilm lines from a FLUENT CFD analysis. (Courtesy of Fluent.)

essary for a basic understanding to run a modern CFD code are discussed in this guide.

Mostly orientated for teaching aeronautical students, this guide approaches the material from the compressible point of view. It can be argued, however, that this view does give more insight into the mathematical and numerical problems in CFD.

1.4 Literature

A more in-depth discussion of the discretisation of the incompressible flow equations with particular focus on heat transfer is given by Tannehill et al. [1]. A wealth of information about the basics of discretisation CFD can be found in the two volumes by Hirsch [2]. Fortified with the reading of this guide, Hirsch's books might prove a bit more easy to navigate. Another recommended text is by Ferziger and Perić [3] dedicated to incompressible flow. The book by Blazek [4] is a very thorough reference discussing the detailed aspects of discretisations for compressible flow. The www also offers a number of short introductory guides. A good starting point is www.cfd-online.com.

1.5 Ingredients

The following basic ingredients are required for a CFD solution:

The physical model embodies our understanding of the physical behaviour. Typical CFD uses the continuum assumption and this book refers to the many textbooks in fluid mechanics that covers this aspect.

The mathematical model translates the flow physics into equations. Chapter 2 recalls how the equations are derived from control volume analysis. The book also introduces a small number of model equations for which an exact solution or behaviour is known. These model problems are used to test the numerical methods.

CFD also introduces additional models to the flow equations e.g., through turbulence modelling. The huge amount of detail that is relevant in turbulent flows, or, if one prefers, the limitations in computational power, requires us to approximate the behaviour of the fine detail with large scale models.

The book does not aim to cover the derivation or the mathematical formulation of these turbulence models. This is left to the scientific literature. The principles of the modelling approaches are presented to the level of detail that a user needs to know in order to design an appropriate mesh or to assess the validity of the solution.

The numerical method or 'discretisation' translates the continuous equations to the computer using a discrete set of points. The book takes the point of view of the user of a CFD software package where the developers have implemented a suitable discretisation of the equations. The user needs to understand the effects of the choices she/he can make when using the software, such as e.g., deciding between first-order and second-order accuracy.

But the aim of the book is not to cover the complete toolbox of numerical analysis. For this we refer to the more advanced texts of [1], [4] or [2].

The mesh defines the location and connectivity of these points, and the user has to be aware of types of meshes and issues with mesh density or mesh quality. The pace of development of mesh generation tools is rapid; hence, the book can only cover the principles.

Pre- and post-processing tools are needed to set up and analyse the results. The flow equations are highly non-linear and we currently do not possess a theory to quantify the errors in our numerical flow simulations.

The user needs to carefully study the results, and apply the limited tools we have to assess the effects of various sources of errors on the quantities of interest that we extract from the solutions, such as e.g., the lift and drag of an aerofoil.

In this analysis the user's knowledge of the flow behaviour and the practice of CFD need to come together. The final two chapters of the book focus on this area.

1.6 Organisation of the chapters

The chapters are organised as follows:

Chapter 2 reviews the underlying physical model and how this leads to the partial differential equations of fluid flow.

Chapter 3 introduces relevant concepts as the finite volume method in a broad way, enabling the reader to start practising with a CFD code.

Chapter 4 looks at the properties of discretisations in more detail and discusses consistency, truncation error and order of accuracy. Sec. 4.3.1 discusses artificial viscosity. Sec. 4.5 discusses the merits of different iterative schemes to solve the discretised systems of equations.

Chapter 5 discusses how to assign boundary conditions.

Chapter 6 introduces the basic theory of modelling of turbulent boundary layers and gives a brief overview over the most popular turbulence models.

Chapter 7 surveys the major grid generation techniques and discusses the influence of grid generation on the solution quality.

Chapter 8 presents the methods used in CFD to assess the accuracy of the solution.

Chapter 9 concludes with a number of case studies. The effects of typical choices in discretisation, meshing, boundary conditions and model choices are analysed and the accuracy of the solutions is assessed.

1.7 Exercises

1.1 Summarise the main advantages of CFD over wind-tunnel testing.

1.2 Summarise the main disadvantages of CFD over wind-tunnel testing.

1.3 What are the main ingredients of a CFD simulation?

2

Physical and mathematical principles of modern CFD

2.1 The physical model

The most basic ingredient to CFD is a physical model of the flow to be simulated. This model embodies our understanding of the physical behaviour of the flow. As will be shown, this model does take some simplifying assumptions which make the model simpler, but also imposes some limitations on its validity.

2.1.1 Continuum assumption

In fluid dynamics we typically are not concerned with the chaotic motion of individual particles of atomic or molecular size. For one, the number of particles is much too large to be tractable; for another, there are so many particles that the impact of an individual particle would not affect the lift or drag of the body that is immersed in the flow.

In effect we assume that there have been so many collisions between particles that we can consider only the statistical average of their bulk flow. This is called the *continuum assumption.*

There are applications which would require to consider individual particles in the flow. These might be flow around space vehicles re-entering the outer atmosphere where there are very few particles. The continuum assumption is also not entirely satisfied in granular flows that carry large particles, or if the motion of red blood cells is of interest in the blood flow in the microvasculature.

Standard CFD techniques rely on the continuum assumption, and flows where this assumption is not valid are disregarded in the following.

2.1.2 Lagrangian vs. Eulerian description

If we analyse the dynamics of a solid object we typically choose to consider the trajectory of the solid object in space. We tend to observe the same element of material (or collection of particles) and apply Newton's laws to them, such as $F = m \cdot a$. This is called the *Lagrangian*[1] viewpoint.

In fluids, however, any particular collection of particles can move around freely and dissociate from each other. Hence it is difficult to "mark" a few particles and follow their trajectories. Instead, we fix an "observation window" in space, the *control volume*, and then observe how the flow streams through it. This is called the *Eulerian*[2] viewpoint. Using this viewpoint simplifies the expression of principles such as mass conservation.

It can be, however, instructive at times to switch viewpoints, e.g., if we want to understand the interaction between pressure and velocity in a flow field, it may help to put ourselves in the place of a particle and see what motion this particle would undergo in this pressure field, or what forces from pressure and friction this particle would need to experience to follow a particular trajectory.

2.1.3 Conservation principles

Once we start viewing the flow as a continuum streaming through the control volume, it becomes natural to consider the balance of the transported quantities in that control volume.

At each location in space we can define a suitably sized control volume and analyse how much mass, momentum, energy (or any other quantity of interest) is entering the control volume and how much is exiting. In the steady state the amount of these quantities in each control volume is constant in time. So what enters minus what exits the control volume has to be balanced by creation or destruction in the control volume.

For example, mass is not created or destroyed; hence, we speak of the *mass conservation equation*. The momentum on the other hand is not always conserved: surface forces such as pressure gradients or wall shear stresses and volume forces such as gravity affect the momentum. We, hence, better refer to the *momentum balance equation*.

2.2 The mathematical model: the equations of fluid flow

Having established a physical model how we view what fluid flow is and the principle of mass conservation and momentum balance, we can now derive the relevant equations.

[1]Joseph-Louis Lagrange, Italian-born mathematician and astronomer who mainly worked in Germany and France, 1736-1813.

[2]Leonhard Euler, Swiss-born mathematician and physicist mainly working in Germany and Russia, 1707-1783.

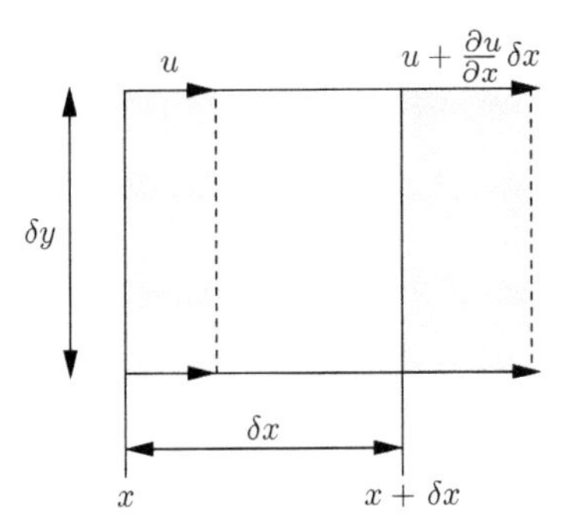

FIGURE 2.1
Mass conservation on a small fluid element in one dimension.

2.2.1 Mass conservation in 1-D

In this section we will briefly recall the derivation of the mass conservation or *continuity* equation in Cartesian coordinates x, y.[3] Consider a fixed rectangular small fluid element having a unit depth in z, as shown in Fig. 2.1. To make things simple, consider the flow to be uniform in z, i.e., no changes in the z direction. But keeping this nominal z-direction in our analysis will maintain the correct dimensions for volumes and areas.

To start with, let's simplify this flow even further and assume also constant flow in y; hence, we only have variations in the x-direction. Let us also assume that the flow is steady, i.e., there are no changes in time. We assume that the fluid element is so small that we can use the first terms of a Taylor[4] expansion to approximate the mass flux $(\rho u)|_{x+\delta x}$ at the right edge of the element $x + \delta x$ in terms of the mass flux $(\rho u)|_x$ at x and its derivatives:

$$(\rho u)|_{x+\delta x} = (\rho u)|_x + \frac{d(\rho u)}{dx}\delta x + \frac{1}{2}\frac{d^2(\rho u)}{dx^2}\delta x^2 + O(\delta x^3)$$

where the term $O(\delta x^3)$ refers to the terms which are third order in δx or higher which have been dropped from the series. The amount of fluid entering on the left at x in time δt is

$$\dot{m}|_x \delta t = (\rho u)|_x \delta y \delta z \delta t;$$

the amount of fluid exiting on the right at $x + \delta x$ in the same time interval is

$$\dot{m}|_{x+\delta x} \delta t = (\rho u)|_{x+\delta x} \delta y \delta z \delta t.$$

[3]The derivation uses *control volumes* and Taylor series expansions. If the reader is not sufficiently familiar with these, any basic fluids textbooks will provide the relevant background.

[4]Brook Taylor, English mathematician, 1685-1731

The change of mass in the control volume in time δt is the difference in mass flow rate entering at x minus the mass flow rate exiting at $x + \delta x$ multiplied by the length of the time interval δt,

$$\begin{aligned} \delta m &= \dot{m}|_x \delta t - \dot{m}|_{x+\delta x} \delta t \\ &= (\rho u)|_x \delta y \delta z \delta t - (\rho u)|_{x+\delta x} \delta y \delta z \delta t. \end{aligned} \tag{2.1}$$

Using the Taylor expansion for ρu we find

$$\begin{aligned} \delta m = & \Big((\rho u)|_x \Big) \delta y \delta z \delta t \\ & - \left((\rho u)|_x + \frac{d(\rho u)}{dx} \delta x + \frac{1}{2} \frac{d^2(\rho u)}{dx^2} \delta x^2 + O(\delta x^3) \right) \delta y \delta z \delta t. \end{aligned}$$

The term $(\rho u)|_x$ appears on both sides and can be subtracted out; similarly, we can divide by the size of the control volume $\delta x \delta y \delta z$ and by the time interval δt to find

$$\frac{\delta m}{\delta x \delta y \delta z \delta t} = \frac{\delta \rho}{\delta t} = -\frac{d(\rho u)}{dx} - \frac{1}{2} \frac{d^2(\rho u)}{dx^2} \delta x + O(\delta x^2). \tag{2.2}$$

(Note that the term $O(\delta x^2)$ means "of the order of", so its sign is undetermined.)

The mass flow gradient $\frac{d(\rho u)}{dx}$ can be arbitrarily large, but if we make the size of our fluid element sufficiently small ($\lim \delta x \to 0$), then $\delta x \ll 1$ and the second and third terms on the right-hand side in (2.2) are negligible,

$$\frac{\delta \rho}{\delta t} = -\frac{d(\rho u)}{dx} = -\rho \frac{du}{dx} - u \frac{d\rho}{dx}. \tag{2.3}$$

However, we have assumed that the flow is steady, so there cannot be any change in the mass in this fluid element, or since the control volume is fixed, the density cannot change in time,

$$\frac{\delta m}{\delta x \delta y \delta z \delta t} = \frac{\delta \rho}{\delta t} = 0.$$

Let us analyse the meaning of (2.3) in more detail and first consider the more general case of a fluid that is compressible, so both velocity u and density ρ can vary in x, but not in time t as we assumed steady flow. In this case we could satisfy Eq. 2.3 if the change in velocity matches the change in density,

$$\frac{du}{dx} = -\frac{u}{\rho} \frac{d\rho}{dx}.$$

For example, if the velocity gradient is negative, i.e., flow is slowing down, then flow enters faster at x than it exits at $x + \delta x$. In this case the density gradient has to be positive, i.e., as the velocity decreases, the density has to increase. In the steady state for this one-dimensional flow the mass flux $(\rho u)|_x$ entering at x has to equal the mass flux $(\rho u)|_{x+\delta x}$ exiting at $x + \delta x$.

In the case of incompressible flow the density is constant, $\rho = \text{const}$, hence the density cannot vary, $\frac{d\rho}{dx} = 0$. As a consequence also $\frac{du}{dx} = 0$, the velocity cannot vary in this one-dimensional case. Only if we allow matching velocity variations in y or z can we have a non-uniform velocity field in the incompressible case: the components of the velocity field are all coupled together through the continuity equation.

2.2.2 Mass conservation in 3-D

We can apply the same analysis as for Eq. 2.1 to 3-D mass conservation where the velocity field will vary in all directions, x, y and z:

$$\begin{aligned} \delta m = \; & \dot{m}|_x \delta t - \dot{m}|_{x+\delta x} \delta t \\ & + \dot{m}|_y \delta t - \dot{m}|_{y+\delta y} \delta t \\ & + \dot{m}|_z \delta t - \dot{m}|_{z+\delta z} \delta t. \end{aligned}$$

Of course, there are now derivatives in all three space directions, so we have to use the partial derivative, but the flux balances in each direction are developed independently of each other as for Eq. 2.1; in particular the terms for $\dot{m}|_x, \dot{m}|_y, \dot{m}|_z$ cancel as before to produce

$$\delta m = -\frac{\partial(\rho u)}{\partial y} \delta x \delta y \delta z \delta t - \frac{\partial(\rho v)}{\partial y} \delta y \delta z \delta x \delta t - \frac{\partial(\rho w)}{\partial z} \delta z \delta x \delta y \delta t. \tag{2.4}$$

Dividing (2.4) also by the size of the control volume $\delta x \delta y \delta z$, and considering small enough time intervals such that the change in density becomes

$$\lim_{\delta t \to 0} \frac{\delta m}{\delta z \delta x \delta y \delta t} = \lim_{\delta t \to 0} \frac{\delta \rho}{\delta t} = \frac{\partial \rho}{\partial t},$$

we obtain the continuity equation for an infinitesimal control volume in 3-D:

$$\frac{\partial \rho}{\partial t} + \frac{\partial(\rho u)}{\partial x} + \frac{\partial(\rho v)}{\partial y} + \frac{\partial(\rho w)}{\partial z} = 0, \tag{2.5}$$

In the case of a compressible fluid we can have more mass flowing into a control volume than flowing out. In this case $\frac{\partial \rho}{\partial t}$ is positive, the density increases with time, and the fluid is compressed.

In the case of an incompressible fluid $\rho = \text{const}$, $\frac{\partial \rho}{\partial t} = 0$. Then Eq. 2.5 simplifies to the *incompressible continuity equation*,

$$\frac{\partial u}{\partial x} + \frac{\partial v}{\partial y} + \frac{\partial w}{\partial z} = 0. \tag{2.6}$$

As we cannot press more mass into a control volume, the inflow has to balance the outflow at any instance at any size control volume, e.g., if we have $\frac{\partial u}{\partial x} < 0$, i.e., flow slows down in the x-direction, more flows enters at x than leaves at $x + \delta x$, then the excess mass flow has to exit in the other two directions.

This is called the *divergence constraint.* Eq. 2.6 is called the *divergence* of the velocity field.

2.2.2.1 Mass conservation: example

Consider a two-dimensional flow field where u, the x-component of velocity, decreases linearly with x along the x-axis:

$$u(x) = -ax.$$

How does v, the y-velocity, behave along the x-axis?

$$\begin{aligned} \frac{\partial u}{\partial x} + \frac{\partial v}{\partial y} &= 0 \\ \frac{\partial v}{\partial y} &= a \\ v &= \int a dy = ay + c. \end{aligned}$$

v has to increase linearly to compensate for the decrease in u.

2.2.2.2 Continuity over finite-size control volumes

In the preceding section we have derived the continuity equation by collecting all the mas fluxes over a finite size control volume in Eq. 2.4. The mass conservation statement is very evident in this viewpoint: the difference between what goes in and what goes out must be accumulated inside. Let us label this form the *integral form* of the continuity equation.

We then shrink that control volume to zero to obtain the *differential* or *strong* form of the continuity equation, Eq. 2.5, which links the velocity gradients in a particular point.

Both forms of the continuity equation are of course valid, provided that the density and velocity fields are not discontinuous but are differentiable.[5]

Both forms are also valid starting points for a numerical approximation of the continuity equation. We will see later that the differential form leads to the finite difference method, while the integral form leads to the finite volume method. It is the latter that is by far the most widely used method for CFD due to its built-in conservation principles. However, in most cases it is much easier to analyse the behaviour of a method in finite difference form, so we shall also use this in our examples.

[5] If they are discontinuous, as e.g., could be the case in supersonic flows with shock waves, we do not have a valid Taylor expansion and the differential form is not valid. But we will limit our discussions here to subsonic or incompressible flows.

2.2.3 Divergence and gradient operators, total derivative

2.2.3.1 Divergence and gradient operators

In this section we recall the basic gradient and divergence operators and their application. Further reading can be found in the basic fluid dynamics textbooks.

In the continuity Eq. 2.4 we observe that we repeat the same terms cyclically for all three dimensions. It is very convenient to use a shorthand, "vector", notation for the gradient $\nabla(a)$ of a scalar variable a,

$$\nabla(a) = \begin{pmatrix} \frac{\partial a}{\partial x} \\ \frac{\partial a}{\partial y} \\ \frac{\partial a}{\partial z} \end{pmatrix}$$

to collect all those terms in a single symbol. The result is a vector.

Applying the gradient operator not to a scalar but to a vector variable such as the velocity field $\vec{u}$ with components u, v, w,[6] we obtain the scalar product of the gradient operator with that vector, which is the divergence. The incompressible continuity equation can then be written

$$\begin{aligned} \nabla \cdot \vec{u} &= \left(\frac{\partial}{\partial x}, \frac{\partial}{\partial y}, \frac{\partial}{\partial z} \right) \begin{pmatrix} u \\ v \\ w \end{pmatrix} \\ &= \frac{\partial u}{\partial x} + \frac{\partial v}{\partial y} + \frac{\partial w}{\partial z} = \operatorname{div}(\vec{u}) = 0. \end{aligned} \tag{2.7}$$

Using the divergence operator in the compressible form of the continuity equation becomes

$$\begin{aligned} \frac{\partial \rho}{\partial t} + \frac{\partial(\rho u)}{\partial x} + \frac{\partial(\rho v)}{\partial y} + \frac{\partial(\rho w)}{\partial z} &= 0 \\ \frac{\partial \rho}{\partial t} + \nabla \cdot (\rho \vec{u}) = \frac{\partial \rho}{\partial t} + \operatorname{div}(\rho \vec{u}) &= 0. \end{aligned} \tag{2.8}$$

2.2.4 The total or material derivative

In the derivation of the momentum equation that follows this section, we will need to express the acceleration of a fluid particle travelling through the flow field.

Using a Lagrangian description (tracking a specific particle), a property S of a fluid particle moving in the flow field depends on time and location.

[6] Note the difference between the entire vector $\vec{u}$ and its x-component u.

A change in this property δS over a small time δt and distance $\delta x, \delta y, \delta z$ is given by

$$\delta S = \frac{\partial S}{\partial t}\delta t + \frac{\partial S}{\partial x}\delta x + \frac{\partial S}{\partial y}\delta y + \frac{\partial S}{\partial z}\delta z. \quad (2.9)$$

We can recognise the contributions to the change δS:

- The time derivative or *local derivative* $\frac{\partial S}{\partial t}$ measures unsteady changes in the flow at a constant location. It is zero in steady flow, but the property S will still change as the particle moves through the flow field.
- The *convective derivative* $u\frac{\partial S}{\partial x} + v\frac{\partial S}{\partial y} + w\frac{\partial S}{\partial z}$ measures the change in S as the particle sweeps through different regions of the flow.

As an example, consider a wind-tunnel that is running at steady speed; let's assume laminar flow and disregard any turbulent fluctuations in the flow. At a particular location x, y, z the speed is constant, and the local derivative $\frac{\partial S}{\partial t}$ is zero. But as the particles move from the wide cross section of the settling area to the narrower test section, the flow speeds up and particles accelerate, and the convective derivative is non-zero, $u\frac{\partial S}{\partial x} + v\frac{\partial S}{\partial y} + w\frac{\partial S}{\partial z} \neq 0$.

Conversely consider incompressible flow in a straight pipe of uniform cross section. If we open the tap we increase the flow speed at every location; the local derivative $\frac{\partial S}{\partial t}$ is non-zero. But since the cross section is invariant, the convective derivative $u\frac{\partial S}{\partial x} + v\frac{\partial S}{\partial y} + w\frac{\partial S}{\partial z}$ is zero as the particles maintain the same speed in all cross sections.

We can write Eq. 2.9 compactly using the vector notation:

$$\frac{DS}{Dt} = \frac{\partial S}{\partial t} + (\vec{u} \cdot \nabla)S.$$

Dividing by δt and taking the limit for $\delta t, x, y, z \to 0$:

$$\frac{\delta S}{\delta t} = \frac{\partial S}{\partial t} + \frac{\partial S}{\partial x}\frac{\partial x}{\partial t} + \frac{\partial S}{\partial y}\frac{\partial y}{\partial t} + \frac{\partial S}{\partial z}\frac{\partial z}{\partial t}.$$

We call this the material derivative DS/Dt:

$$\frac{DS}{Dt} = \frac{\partial S}{\partial t} + u\frac{\partial S}{\partial x} + v\frac{\partial S}{\partial y} + w\frac{\partial S}{\partial z}.$$

If apply the total derivative to each of the velocity components in turn, we obtain the acceleration of the particle in the flow field, or the total derivative of the velocity:

$$\frac{Du}{Dt} = \frac{du}{dt} = \frac{\partial u}{\partial t} + u\frac{\partial u}{\partial x} + v\frac{\partial u}{\partial y} + w\frac{\partial u}{\partial z}$$
$$\frac{Dv}{Dt} = \frac{dv}{dt} = \frac{\partial v}{\partial t} + u\frac{\partial v}{\partial x} + v\frac{\partial v}{\partial y} + w\frac{\partial v}{\partial z}$$
$$\frac{Dw}{Dt} = \frac{dw}{dt} = \frac{\partial w}{\partial t} + u\frac{\partial w}{\partial x} + v\frac{\partial w}{\partial y} + w\frac{\partial w}{\partial z}.$$

2.2.5 The divergence form of the total derivative

Let us recall that we obtained the steady continuity equation in differential form. Eg. 2.8, by summing the mass fluxes over the surfaces of a control volume. We can of course do the reverse and integrate the divergence of the velocity $\vec{u}$ over our control volume.

In our derivation in Sec. 2.2.2 we had specifically aligned the faces of our rectangular control volume to be orthogonal to the coordinate directions. In the general case we need to consider arbitrary face orientations where the velocity component and the face are not perpendicular to each other. The mass flux that changes the mass balance in the control volume is the mass that is transported through the face, not parallel to it. The component of the velocity $u_\perp$ normal to the face is obtained with the scalar (or dot) product,

$$(\rho u)_\perp = \rho \vec{u} \cdot \vec{n}, \tag{2.10}$$

where $\vec{n}$ is the unit normal of the face. Integrating the mass fluxes normal to the control volume accumulates the net mass flow rate in to or out of the control volume, which is equal to its rate of change of mass. We have hence obtained the *integral* form of the continuity equation,

$$\iiint_V \frac{\partial \rho}{\partial t}\, dV + \oiint_S \rho \vec{u} \cdot \vec{n}\, dS = 0, \tag{2.11}$$

where V is the size of the control volume, S is its surface and $\vec{n}$ is the unit normal to the surface. Comparing (2.11) to (2.8) we find that the two versions of the continuity equations are equivalent

$$\iiint_V \left(\frac{\partial \rho}{\partial t} + \operatorname{div}(\rho \vec{u}) \right) dV = \iiint_V \frac{\partial \rho}{\partial t}\, dV + \oiint_S \rho \vec{u} \cdot \vec{n}\, dS = 0. \tag{2.12}$$

In particular,

$$\iiint_V \operatorname{div}(\rho \vec{u})\, dV = \oiint_S \rho \vec{u} \cdot \vec{n}\, dS$$

which is, of course, simply Gauss' divergence theorem, and we could have used that directly on the vector field $\rho \vec{u}$ to obtain the integral or divergence form of the continuity equation in the steady case. For incompressible flow this reduces to

$$\iiint_V \operatorname{div}(\vec{u})\, dV = \oiint_S \vec{u} \cdot \vec{n}\, dS = 0. \tag{2.13}$$

The finite volume method uses this integral form to approximate the continuity equation over small discrete control volumes.

2.2.6 Reynolds' transport theorem

The integral form of the continuity equation (2.11) can be generalised to the transport of any quantity that is conserved under transport. While mass is exactly conserved, other quantities may only be 'balanced'. Consider e.g., the energy balance. Transport or *advection* itself can not create or destroy energy in the control volume. If there was only transport by the flow field we can work out how much is accumulated in a control volume by looking at the difference between what is transported in compared to what is transported out. However, the energy balance in the control volume may be affected by generation of heat due to friction or heat conduction at the wall. We have to account for these source terms separately, but we can derive a generic conservative formulation for the transport.

In the following we shall use primarily *intensive* quantities such as density ρ or velocity u which are independent of the size of the control volume or the size of the fluid element. On the other hand it is the *extensive* quantities such as mass m or momentum mu that are conserved. The amount of the conserved, extensive quantity Φ in a fixed control volume V can then be determined by integrating the product of density and the intensive quantity,

$$\Phi = \iiint_V \rho\phi dV. \tag{2.14}$$

Selecting e.g., $\phi = 1$ then produces mass; its transport conservation statement, the continuity equation, $\phi = u$, produces the x-momentum transport equation. Introducing ϕ into (2.11), the advective conservation of ϕ that is transported by the flow field is then given by

$$\frac{\partial \rho\phi}{\partial t} + \operatorname{div}(\rho\phi\vec{u}). \tag{2.15}$$

Note the absence of an equal sign, since the terms in (2.15) only represent the advective terms in the balance statement. Before we add the source terms to this to derive the complete balance of Φ, let us derive an alternative form that is valid for incompressible flow.

Using the product rule for ϕ and ρ in the divergence term one finds

$$\begin{aligned} \frac{\partial \rho\phi}{\partial t} + \operatorname{div}(\rho\phi\vec{u}) &= 0 \\ \rho\frac{\partial \phi}{\partial t} + \phi\frac{\partial \rho}{\partial t} + \rho \operatorname{div}(\phi\vec{u}) + \phi \operatorname{div}(\rho\vec{u}) &= 0 \\ \rho\frac{\partial \phi}{\partial t} + \rho \operatorname{div}(\phi\vec{u}) + \phi\left(\frac{\partial \rho}{\partial t} + \operatorname{div}(\rho\vec{u})\right) &= 0. \end{aligned}$$

The flow field does satisfy continuity (2.8); hence, the term in the parentheses on the right vanishes,

$$\rho\frac{\partial \phi}{\partial t} + \rho \operatorname{div}(\phi\vec{u}) = 0 \tag{2.16}$$

This form is called the *conservative*, *weak* or *divergence* form; we shall mainly use the label *conservative* form in this book. Its conservation properties become apparent if, as previously shown for the continuity equation, we integrate over the control volume and use Gauss' divergence theorem to obtain

$$\iiint\limits_V \left(\rho \frac{\partial \phi}{\partial t} + \rho \operatorname{div}(\phi \vec{u}) \right) dV = \iiint\limits_V \rho \frac{\partial \phi}{\partial t} dV + \oiint\limits_S \phi \vec{u} \cdot \vec{n}\, dS, \tag{2.17}$$

i.e., the rate of change of ϕ is driven by the balance of ϕ-fluxes.

Applying the product rule to the divergence term in (2.16) we obtain

$$\rho \frac{\partial \phi}{\partial t} + \rho \phi \operatorname{div}(\vec{u}) + \rho \vec{u}\, \nabla \phi.$$

If we consider incompressible flow, with $\operatorname{div} \vec{u} = 0$, (2.7), then the second term vanishes and we obtain

$$\rho \left(\frac{\partial \phi}{\partial t} + \vec{u}\, \nabla \phi \right) = \rho \frac{D\phi}{Dt}, \tag{2.18}$$

or, in component form,

$$\rho \left(\frac{\partial \phi}{\partial t} + u \frac{\partial \phi}{\partial x} + v \frac{\partial \phi}{\partial y} + w \frac{\partial \phi}{\partial z} \right)$$

which is the density-weighted total or material derivative of ϕ.

This form is called the *non-conservative form.* This form lends itself to simple discretisations and we will use it for some of our example problems. But as the name indicates, it is not guaranteed to be conservative, so it is not the first choice for CFD.

2.2.7 Transport of a passive scalar

The Eqs. 2.16 and 2.18 describe how some passive scalar quantity[7] changes with time t and along distance x given the fixed speed a. An example of this behaviour would be the advection of spot of dye in a river, the value u being the concentration of dye which does not change the speed of flow in the river. Note however that in the advection equation there we have also taken out the viscous or diffusive term, i.e., in our simplified model the spot of dye would not spread and diffuse as it is being transported downriver.

2.3 The momentum equations

The momentum equations for a fluid can be derived by considering the force balance on a moving particle. Force being a vector quantity, we expect three momentum equations, one for each of the coordinate directions.

[7]Passive in the sense that it does not affect the flow field. The speed a remains constant regardless of the value of u.

You will be familiar with Newton's second law, $F = m\,a$, which relates the sum of forces F to the acceleration a of the mass m. Since it is difficult to track the motion of a specific ensemble of particles forming the mass m, we prefer to use the Eulerian viewpoint and consider the more general form of Newton's law for all three coordinate directions:

$$\frac{d(m\vec{u})}{dt} = \vec{F} \tag{2.19}$$

where $\vec{u}$ is the velocity and $\vec{F}$ is the force vector.[8] To emphasize the conservation of momentum, let us integrate Newton's law over a fixed control volume, using Reynolds' transport theorem (2.17), to obtain

$$\iiint\limits_V \frac{d(m\vec{u})}{dt} dV = \iiint\limits_V \rho \frac{\partial \vec{u}}{\partial t} dV + \oiint\limits_S \rho \vec{u}\vec{u}\cdot\vec{n}\, dS, = \iiint\limits_V \vec{F} dV. \tag{2.20}$$

We can see that the rate of change of momentum is equal to the net momentum transport and the sum of the forces that are applied.

2.3.1 Examples of momentum balance

2.3.1.1 Boundary layers

In the next two sections we will derive the inviscid and the viscous momentum equation. We shall be starting with the inviscid momentum equations, the *Euler equations*, and then extend the analysis to include viscous terms to obtain the viscous momentum equations, or *Navier-Stokes equations*.[9] We will limit the derivation to incompressible flows, derivations that include compressible flows can be found in the basic fluid dynamics textbooks.

2.3.2 The inviscid momentum equation — the Euler equation

In Sec. 2.3 we have derived a generic integral expression for the momentum balance (2.20). What remains to be done for the Euler equations is to calculate the sum of forces on the right-hand side. In the case of inviscid flow we neglect the effect of viscous shear forces; fluid elements can slide past each other without resistance, but we do need to account for pressure forces and gravity. Fig. 2.2 shows the pressure forces on the faces of fluid element. Recall that pressure is a scalar. It has no particular direction. It only acquires a direction when acting on a surface. The resulting force is then perpendicular

[8] Of course, in a Lagrangian viewpoint the fixed mass can be moved out of the derivative to obtain $F = m\,a$.

[9] Many authors, including this one, consider the Navier-Stokes equation to be the set of continuity and viscous momentum equations, while some only use the term to refer to the viscous momentum equations.

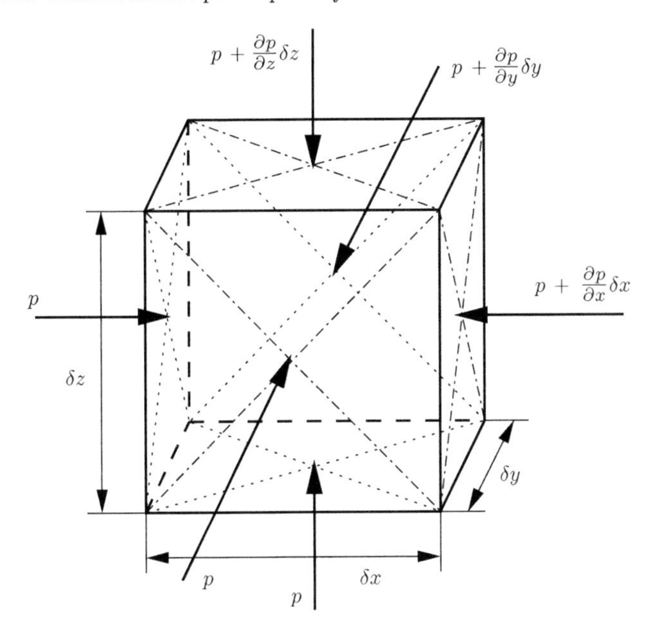

FIGURE 2.2
Pressure forces on a fluid element.

to the surface. Here we take a positive pressure as being compressive, i.e., the pressure on the face at x with normal pointing in the $-x$-direction results in positive force in the x-direction, while the pressure on the face at $x + \delta x$ results in a negative force.

Using Taylor expansion to first order (retaining only the linear term) as we have done for the continuity equation in Sec. 2.2.1, we can determine the net pressure force in x to be

$$\begin{aligned} F_{p,x} &= \left(p|_x - p|_{x+\delta x}\right)\delta y \delta z \\ &= \left(p|_x - \left(p|_x + \frac{\partial p}{\partial x}|_x \delta x\right)\right)\delta y \delta z \\ &= -\frac{\partial p}{\partial x}\delta x \delta y \delta z = -\frac{\partial p}{\partial x}\delta V \end{aligned}$$

where δV is the size of the control volume $\delta x \delta y \delta z$. Similarly we obtain in the y- and z-directions

$$\begin{aligned} F_{p,y} &= -\frac{\partial p}{\partial y}\delta V \\ F_{p,z} &= -\frac{\partial p}{\partial z}\delta V. \end{aligned}$$

In vector notation the pressure force vector is

$$F_P = \begin{bmatrix} F_{p,x} \\ F_{p,y} \\ F_{p,z} \end{bmatrix}.$$

In the Euler equations the flow is considered inviscid, so there are no shear forces on the volume surface. The fluid will e.g., be subject to gravity (although in gases typically gravitational effects are negligible). Denoting the direction of the sum of body forces as $\vec{b}$, e.g., with $\vec{b} = [0, 0, -g]^T$, if the vertical is aligned with the z-axis and g is the graviational acceleration, we have

$$F_{b,z} = \rho \vec{b} \delta V.$$

Both pressure and body forces are proportional to the size of the control volume δV. The sum of forces per volume is then

$$f = \frac{1}{\delta V} (F_P + F_G) = \nabla p + \rho \vec{b}. \tag{2.21}$$

The integral form of the momentum equation then becomes

$$\iiint_V \rho \frac{\partial \vec{u}}{\partial t} dV + \oiint_S \rho \vec{u} \vec{u} \cdot \vec{n} \, dS, = \iiint_V \nabla p + \rho \vec{b} \, dV. \tag{2.22}$$

This integral form is *conservative* form of the momentum equation. It is this form that will be used in CFD. However, it is a bit unwieldy for simple analysis, so similar to what we have done with Reynolds' transport theorem, Eq. 2.18, we can turn the divergence div $\rho \vec{v} \vec{v}$ into the convective derivative $\vec{u} \nabla u$, using continuity and assuming incompressible flow. We then obtain the inviscid momentum equations in differential form and in component notation:

$$\frac{\partial u}{\partial t} + u \frac{\partial u}{\partial x} + v \frac{\partial u}{\partial y} + w \frac{\partial u}{\partial z} + \frac{1}{\rho} \frac{\partial p}{\partial x} + b_x = 0, \tag{2.23}$$

$$\frac{\partial v}{\partial t} + u \frac{\partial v}{\partial x} + v \frac{\partial v}{\partial y} + w \frac{\partial v}{\partial z} + \frac{1}{\rho} \frac{\partial p}{\partial y} + b_y = 0, \tag{2.24}$$

$$\frac{\partial w}{\partial t} + u \frac{\partial w}{\partial x} + v \frac{\partial w}{\partial y} + w \frac{\partial w}{\partial z} + \frac{1}{\rho} \frac{\partial p}{\partial z} + b_z = 0. \tag{2.25}$$

In vector notation, using the total derivative, one obtains

$$\frac{D \vec{u}}{Dt} + \frac{1}{\rho} \nabla p - \vec{f_b} = 0.$$

Even though all practical fluids have viscosity, the Euler equations are actually surprisingly useful. In flows over vehicles or aircraft strong viscous effects are often only present in very thin layers near solid walls, the *boundary layers*. Outside those layers viscous effects are often small. Since the boundary layers are often thin, they do not affect the outer flow significantly and if the flow remains attached, the pressure distribution over the vehicle can be often predicted with acceptable accuracy even when the boundary layers are neglected. However, the Euler model can of course not correctly model viscous effects such as skin friction drag or flow separation.

2.3.3 The viscous momentum equations — Navier-Stokes equations

To extend the Euler equations to include viscous effects, we need to include the shear forces due to viscosity. We shall not derive the stress tensor τ_{ij} from first principles here. The reader is referred to basic fluid dynamics textbooks.

The sum of the forces in the x-direction as shown in Fig. 2.3 can be obtained by summing normal and shear stresses on the two opposing faces. As

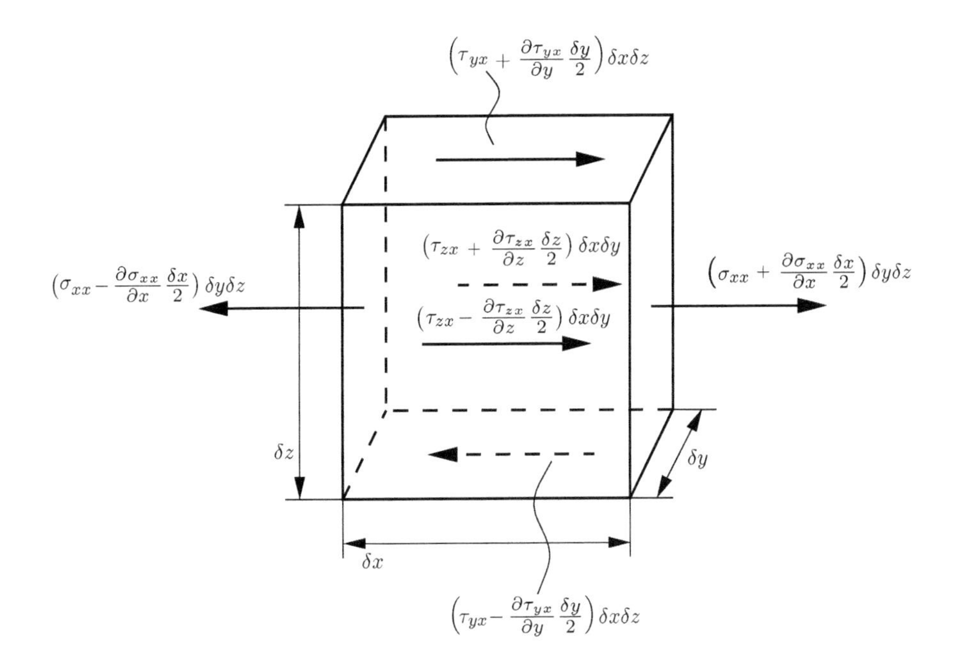

FIGURE 2.3
Stresses in the x-direction on a fluid element.

typical, we use Taylor expansion:

$$
\begin{aligned}
F_x = & \left(-\sigma_{xx} + (\sigma_{xx} + \frac{\partial \sigma_{xx}}{\partial x}\delta x)\right)\delta y \delta z \\
& + \left(-\tau_{yx} + (\tau_{yx} + \frac{\partial \tau_{yx}}{\partial y}\delta y)\right)\delta z \delta x \\
& + \left(-\tau_{zx} + (\tau_{zx} + \frac{\partial \tau_{zx}}{\partial z}\delta z)\right)\delta x \delta y.
\end{aligned}
$$

Dividing by the volume δV produces the sum of surface forces per volume $f_x = F_x/\delta V$ in x:

$$
f_x = \frac{\partial \sigma_{xx}}{\partial x} + \frac{\partial \tau_{yx}}{\partial y} + \frac{\partial \tau_{zx}}{\partial z}. \tag{2.26}
$$

The relations for the y- and z-directions follow straightforwardly. For a Newtonian fluid, where the shear stress is proportional to the shear strain, we have then[10]

$$
\tau_{xy} = \mu\left(\frac{\partial v}{\partial x} + \frac{\partial u}{\partial y}\right).
$$

The rate of strain tensor ϵ_{ij} is defined as

$$
\epsilon_{ij} = \begin{bmatrix}
\frac{1}{2}\left(\frac{\partial u}{\partial x} + \frac{\partial u}{\partial x}\right) & \frac{1}{2}\left(\frac{\partial v}{\partial x} + \frac{\partial u}{\partial y}\right) & \frac{1}{2}\left(\frac{\partial w}{\partial x} + \frac{\partial u}{\partial z}\right) \\
\frac{1}{2}\left(\frac{\partial u}{\partial y} + \frac{\partial v}{\partial x}\right) & \frac{1}{2}\left(\frac{\partial v}{\partial y} + \frac{\partial v}{\partial y}\right) & \frac{1}{2}\left(\frac{\partial w}{\partial y} + \frac{\partial v}{\partial z}\right) \\
\frac{1}{2}\left(\frac{\partial u}{\partial z} + \frac{\partial w}{\partial x}\right) & \frac{1}{2}\left(\frac{\partial v}{\partial z} + \frac{\partial w}{\partial y}\right) & \frac{1}{2}\left(\frac{\partial w}{\partial z} + \frac{\partial w}{\partial z}\right)
\end{bmatrix}. \tag{2.27}
$$

Using this notation, the stress tensor becomes

$$
\tau_{ij} = 2\mu\epsilon_{ij}, \tag{2.28}
$$

e.g., in 2-D, the shear stress in the xy plane is

$$
\tau_{xy} = 2\mu\epsilon_{ij} = 2\mu\frac{1}{2}\left(\frac{\partial v}{\partial x} + \frac{\partial u}{\partial y}\right) = \mu\left(\frac{\partial v}{\partial x} + \frac{\partial u}{\partial y}\right).
$$

This is a somewhat simplified derivation. Other terms with additional viscosities arise in the general derivation, but those are zero for incompressible flow. Any elementary fluid textbook can be consulted for details for the compressible case. In addition to the shear stresses, we can observe that there is a normal viscous stress on the diagonal of the stress tensor:

$$
\tau_{xx} = 2\mu\frac{\partial u}{\partial x}.
$$

Viscous forces arise due to exchanges of momentum when particles collide

[10]We shall leave the derivation of the stress-strain relationship to basic textbooks.

while in (thermal, Brownian) motion. During those random collisions all momentum components are exchanged, tangential as well as normal momentum. The normal stress σ_{xx} is then composed of the pressure $-p$ (negative as compressive) and the normal viscous stress τ_{xx}. We can also observe that in all cases the stress tensor is symmetric: $\tau_{xy} = \tau_{yx}$.

Having derived the stresses in Eqs. 2.27 and 2.28 we can now assemble the sum of surface forces in the x-direction for the incompressible case for Eq. 2.26, repeated here:

$$f_x = \frac{\partial \sigma_{xx}}{\partial x} + \frac{\partial \tau_{yx}}{\partial y} + \frac{\partial \tau_{zx}}{\partial z}.$$

Using the terms from the strain rate tensor ϵ_{ij}:

$$f_x = \frac{\partial}{\partial x}\left(-p + 2\mu\frac{\partial u}{\partial x}\right) + \frac{\partial}{\partial y}\mu\left(\frac{\partial v}{\partial x} + \frac{\partial u}{\partial y}\right) + \frac{\partial}{\partial z}\mu\left(\frac{\partial w}{\partial x} + \frac{\partial u}{\partial z}\right).$$

Rearranging the terms produces

$$\begin{aligned} f_x &= -\frac{\partial p}{\partial x} + \mu\left(\frac{\partial}{\partial x}\frac{\partial u}{\partial x} + \frac{\partial}{\partial x}\frac{\partial u}{\partial x} + \frac{\partial}{\partial y}\frac{\partial u}{\partial y} + \frac{\partial}{\partial y}\frac{\partial v}{\partial x} + \frac{\partial}{\partial z}\frac{\partial u}{\partial z} + \frac{\partial}{\partial z}\frac{\partial w}{\partial x}\right) \\ &= -\frac{\partial p}{\partial x} + \mu\left(\frac{\partial}{\partial x}\frac{\partial u}{\partial x} + \frac{\partial}{\partial y}\frac{\partial u}{\partial y} + \frac{\partial}{\partial z}\frac{\partial u}{\partial z}\right) + \mu\frac{\partial}{\partial x}\left(\frac{\partial u}{\partial x} + \frac{\partial v}{\partial y} + \frac{\partial w}{\partial z}\right). \end{aligned}$$

The sum in the bracket on the right is zero due to continuity, hence

$$f_x = -\frac{\partial p}{\partial x} + \mu\left(\frac{\partial^2 u}{\partial x^2} + \frac{\partial^2 u}{\partial y^2} + \frac{\partial^2 u}{\partial z^2}\right).$$

Adding the viscous forces to the inviscid momentum equation (2.22) in integral form with surface integral for the convective terms, we find the integral form of the viscous momentum equation:

$$\iiint_V \rho\frac{\partial \vec{u}}{\partial t} dV + \oiint_S \rho\vec{u}\vec{u}\cdot\vec{n}\, dS, = \iiint_V \nabla p + \rho\vec{b}\, dV + \oiint_S \mu\nabla\vec{v}\, dV. \tag{2.29}$$

2.3.4 The incompressible Navier-Stokes equations

Assuming incompressible flow simplifies the equations. Collecting all the terms we obtain in the x-direction:

$$\frac{Du}{dt} = -\frac{1}{\rho}\frac{\partial p}{\partial x} + \frac{\mu}{\rho}\left(\frac{\partial^2 u}{\partial x^2} + \frac{\partial^2 u}{\partial y^2} + \frac{\partial^2 u}{\partial z^2}\right)$$

or, spelling out the total derivative and bringing pressure on the left-hand side:

$$\frac{\partial u}{\partial t} + u\frac{\partial u}{\partial x} + v\frac{\partial u}{\partial y} + w\frac{\partial u}{\partial z} + \frac{1}{\rho}\frac{\partial p}{\partial x} = \nu\left(\frac{\partial^2 u}{\partial x^2} + \frac{\partial^2 u}{\partial y^2} + \frac{\partial^2 u}{\partial z^2}\right).$$

In vector form (still only x-direction, no gravity):

$$\frac{\partial u}{\partial t} + \vec{u}\,\nabla\cdot u + \frac{1}{\rho}\nabla p = \nu\nabla^2 u.$$

Applying the same derivation to the other directions, assuming for simplicity that the only body forces are gravity in the $-z$-direction:

$$\frac{\partial u}{\partial t} + u\frac{\partial u}{\partial x} + v\frac{\partial u}{\partial y} + w\frac{\partial u}{\partial z} + \frac{1}{\rho}\frac{\partial p}{\partial x} + b_x = \nu\left(\frac{\partial^2 u}{\partial x^2} + \frac{\partial^2 u}{\partial y^2} + \frac{\partial^2 u}{\partial z^2}\right) \tag{2.30}$$

$$\frac{\partial v}{\partial t} + u\frac{\partial v}{\partial x} + v\frac{\partial v}{\partial y} + w\frac{\partial v}{\partial z} + \frac{1}{\rho}\frac{\partial p}{\partial y} + b_y = \nu\left(\frac{\partial^2 v}{\partial x^2} + \frac{\partial^2 v}{\partial y^2} + \frac{\partial^2 v}{\partial z^2}\right) \tag{2.31}$$

$$\frac{\partial w}{\partial t} + u\frac{\partial w}{\partial x} + v\frac{\partial w}{\partial y} + w\frac{\partial w}{\partial z} + \frac{1}{\rho}\frac{\partial p}{\partial z} + b_z = \nu\left(\frac{\partial^2 w}{\partial x^2} + \frac{\partial^2 w}{\partial y^2} + \frac{\partial^2 w}{\partial z^2}\right). \tag{2.32}$$

In vector form we have:

$$\frac{\partial \vec{u}}{\partial t} + \vec{u}\,\nabla\cdot\vec{u} + \frac{1}{\rho}\nabla p - \vec{f_b} = \nu\nabla^2\vec{u}. \tag{2.33}$$

2.3.5 Energy balance

One might imagine that the complete description of the flow does always require to include an equation that describes the energy balance in the control volume. Actually, an energy equation is not always required. As you may remember from elementary fluid dynamics, the Bernoulli equation

$$p_{static} + \frac{1}{2}\rho u^2 = const = p_{total} \tag{2.34}$$

is derived from integrating the momentum equation along a streamline in inviscid incompressible flow. Hence if our flow satisfies the momentum equation, it will also respect Bernoulli's equation.[11]

However, in the case of compressible flow we need to distinguish between energy that can be converted back and forth between pressure and kinetic energy and the part of thermal energy that cannot be converted. An energy equation is required for compressible flows.

Even in incompressible flows we may at times need an energy equation. Consider e.g., buoyancy effects in a heat exchanger; the heated walls heat up the fluid which as a consequence expands in volume, i.e., has a reduced density. Hence we need to solve an energy equation to keep track of the temperature in a control volume to be able to determine its density and correctly compute volume forces due to buoyancy. In any cases where the temperature has an interaction with the flow field we will need to solve an energy equation.

[11] 'Respect' in the sense that Bernoulli's equation is only valid in inviscid flow. In viscous flow we have to take energy losses due to friction into account.

2.3.6 Summary of properties for the Navier-Stokes equations

Modern CFD packages solve the 3-D Navier-Stokes equations. Their properties are summarised below. They

- are based on the continuum assumption,
- are a coupled system of 5 non-linear partial differential equations (4 in 2-D) in space and time, closed by an equation of state,
- describe balance of mass, momentum and energy in fluid flow,
- describe the advection and diffusion of a passive scalar (one that does not influence the flow field itself such as e.g., the advection of a spot of dye with the flow),
- fully describe viscous effects including turbulence (as also in turbulent flow mass is conserved, momentum and energy are balanced),
- in their compressible form: describe wave propagation phenomena (such as sound waves), damped in time,
- in their incompressible form: assume an infinitely fast propagation speed of sound waves,
- do not have a known closed-form general solution, so we need to approximate the solution numerically.

In many industries commercially licensed solvers are used such as the Fluent and CFX solvers from Ansys, the Star-CCM+ solver from CD-Adapco or the ACE+ solver from ESI. The open-source solver OpenFOAM is also widely used in industry and freely available. In the aerospace industry many companies and research labs have developed and maintain their own CFD codes to suit specialist needs.

To reduce the computational cost (use of computer runtime and memory) it may be useful to neglect effects of viscosity. In this case the Navier-Stokes equations reduce to the Euler equations and the viscous term in the momentum equation (2.20) is multiplied with zero viscosity μ and vanishes.[12] This form has been used extensively in aerofoil design where boundary layers are very thin and good approximations to the pressure field can be obtained even when viscous effects are neglected. Any viscous effects present in solutions to the Euler equations are an error as the solution should not exhibit any viscosity. We can use the inviscid form of the equations also to identify any errors due to *artificial viscosity* (see Sec. 4.3.1), as no viscous effect should be present in solutions to the Euler equations.

[12]The boundary condition at solid walls needs also to be changed from no-slip to slip.

2.4 Simplified model equations

The Euler and Navier-Stokes equations are a rather complicated set of equations to deal with. To understand the underlying principles, we shall first have a look at three equations which can be seen as simplifications of the flow equations, namely the advection equation, the inviscid Burgers' equation and the heat equation.

2.4.1 The linear advection equation

The momentum equation (2.33) takes external shear and pressure forces into account. As a first simplification, let us for the time being drop these terms and consider flow without viscosity and without a forcing pressure gradient. Additionally, let us simplify the advection (transport) term. In the momentum equation it is a non-linear term as the moment ρu is advected with speed u. Let us make this term linear by fixing the advection speed to be a constant a. This corresponds to considering the property u to be a passive scalar that is transported on fixed velocity field, $u = \phi$, in Reynolds' transport theorem. Finally, let us only consider transport in one dimension, x.

In differential form, the *linear advection equation* is

$$\frac{\partial u}{\partial t} + a\frac{\partial u}{\partial x} = 0, \tag{2.35}$$

In conservative or weak form

$$\frac{\partial u}{\partial t} + \frac{\partial f}{\partial x} = 0 \tag{2.36}$$

where the flux of u is $f = au$.

The linear advection equation describes the transport of a passive scalar neglecting any viscous dissipation. This behaviour is shown in Fig. 2.4 where an initial profile of the scalar u is being advected to the right. In a time Δt the profile will travel $\Delta x = a\Delta t$. The constant advection speed a implies that the solution is constant along lines of $x/t = a$, or that the solution is a function of a single parameter $x - at$: $u = u(x - at)$. Inserted into the advection equation we obtain

$$\begin{aligned} 0 =& \frac{\partial u}{\partial t} + a\frac{\partial u}{\partial x} \\ =& -au + au = 0 \end{aligned}$$

which confirms that $u = u(x - at)$ is a solution to the advection equation.

The linear advection equation can also be formulated in two or three dimensions. Extending the one-dimensional equations to vector notation we

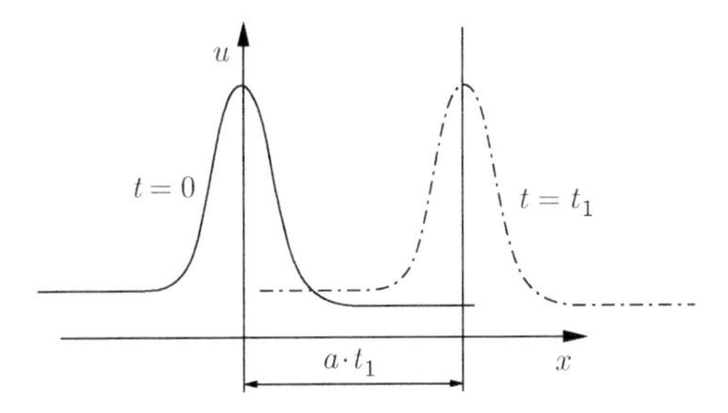

FIGURE 2.4
Advection of a pulse.

obtain in differential form

$$\frac{\partial u}{\partial t} + \mathbf{a}\nabla u = 0,$$
$$\frac{\partial u}{\partial t} + a_x\frac{\partial u}{\partial x} + a_y\frac{\partial u}{\partial y} + a_z\frac{\partial u}{\partial z} = 0, \tag{2.37}$$

where a_x, a_y, a_z are the components of $\mathbf{a}$. The integral or weak form, obtained similarly to (2.14),

$$\iint_A \frac{\partial u}{\partial t} + \oiint \mathbf{a}u\cdot\mathbf{n}\, dS = 0 \tag{2.38}$$

reduces to

$$\frac{\partial u}{\partial t} + \nabla\cdot(\mathbf{a}u) = 0$$

for an infinitesimal control volume.

2.4.2 Inviscid Burgers' equation

The momentum equations have a non-linear term $u\nabla u$; in the linear advection equation we considered a fixed velocity field for the transport, and the convection term became $a\nabla u$. But it is actually the discretisation of the non-linear behaviour of the convection term that needs to be done carefully, or otherwise the discretisation may not be conservative.

To study non-linear advection we will consider the invisicid Burgers'[13] equation,[14]

$$\frac{\partial u}{\partial t} + u\frac{\partial u}{\partial x} = 0 \tag{2.39}$$

[13] Johannes (Jan) Martinus Burgers, Dutch physicist, 1895-1981.

[14] Burgers' original equation also has a viscous term, which we are not considering here. The wikpedia article on Burgers' equation offers a number of interesting animations that show the behaviour of the solution.

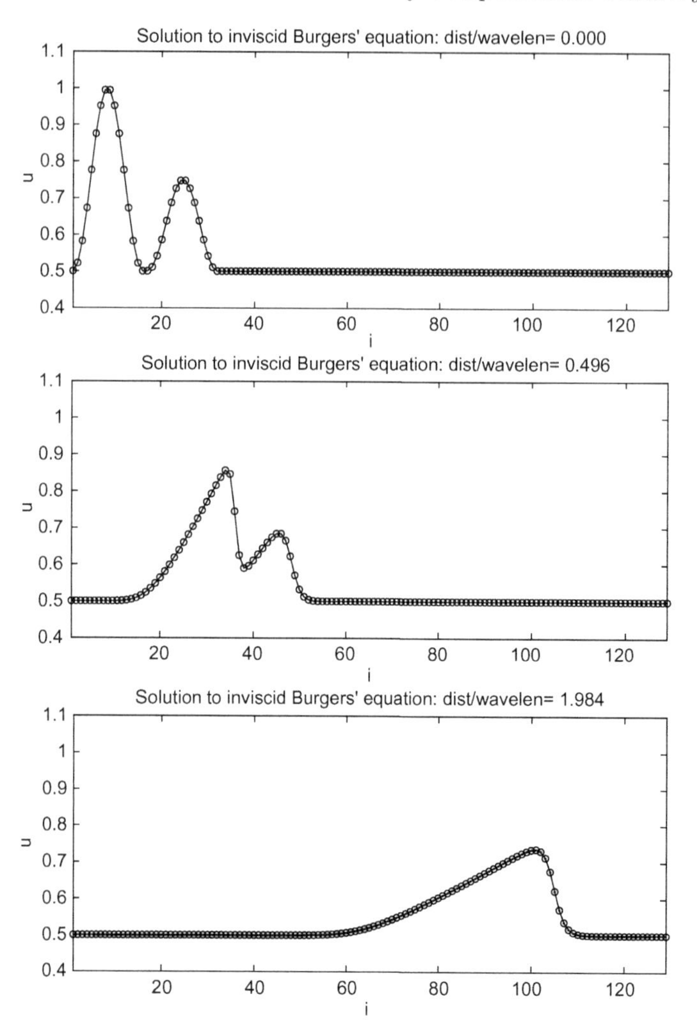

FIGURE 2.5
Solution to Burgers' equation, the crest of the left wave with the higher peak travels faster and swallows the slower wave on the right, resulting in the formation of a steep shock wave.

or in conservative form

$$\frac{\partial u}{\partial t} + \frac{\partial f}{\partial x} = 0 \quad \text{with} \quad f = \frac{1}{2}u^2. \tag{2.40}$$

Similar to the linear advection equation Burgers' equation also describes transport. But the speed of transport of u is proportional to the value of u, just as in the momentum equation.

Consider e.g., a solution with two 'humps' or 'waves' as shown in Fig. 2.5.

The wave on the left has a higher crest and travels faster; it catches up with the slower wave and 'swallows' it. The wave itself steepens, as its own crest travels faster than the right flank or its right foot, and a *shock wave* or wave front will form. The same behaviour is observed in water surface waves: taller waves are faster, and if tall enough they can break. The behaviour that the wave speed is not constant but depends on the flow properties is called *dispersive.*

2.4.3 The heat equation

The heat equation describes the exchange of a property by pure diffusion. One example is the mixing of particles with different temperatures, hence the name heat equation. A diffusive heat flux f is governed by Fourier's law[15]

$$f = -k\nabla T \tag{2.41}$$

where k is a diffusion or heat conduction coefficient and T is the temperature. The negative sign implies that thermal energy flows from hot to cold. Thermal energy is conserved, so integrating the heat flux over the surface of the control volume is equal to the rate of change of thermal energy $Q = c_p \rho T$ in the control volume, where c_P is the specific heat capacity,

$$\frac{\partial Q}{\partial t} + \iiint_V k\nabla T \cdot \vec{n}\, dS = 0. \tag{2.42}$$

The derivation follows the steps as the continuity equation (2.11). Shrinking the control volume V to zero we obtain the differential form of the heat equation,

$$\frac{\partial T}{\partial t} + \frac{k}{c_p \rho}\frac{\partial^2 T}{\partial x^2} = 0. \tag{2.43}$$

In the steady state, once all thermal fluctuations are in equilibrium, we obtain

$$\nabla^2 T = \frac{\partial^2 T}{\partial x^2} + \frac{\partial^2 T}{\partial y^2} + \frac{\partial^2 T}{\partial z^2} = 0 \tag{2.44}$$

which is also called *Laplace's equation.*[16] The equation can also be obtained from the momentum equation by removing convection and pressure gradients, and considering only the steady case. In effect, this removes all terms on the left-hand side of the Navier-Stokes momentum equation (2.29) The only physical mechanism left is pure diffusion of momentum.

Our interest in this equation is however not so much the viscous term in the momentum equation, but the *pressure correction* equation which arises in the discretisation of the system of the incompressible Navier-Stokes eq, (see Sec. 3.6.6). In this case the right-hand side of the equation is not zero, but

[15] Jean Baptiste Joseph Fourier, French mathematician and physicist, 1768-1830.

[16] Pierre-Simon de Laplace, French mathematician and astronomer, 1749-1827.

has a source term, similar to e.g., a chemical reaction generating heat in the field. This type of equation is called a *Poisson* equation.[17] Considering the unknown pressure p and labelling the source term as d, the equation reads

$$\nabla^2 p = d. \tag{2.45}$$

Using a component notation for the gradient operator ∇ in three dimensions, one finds

$$\frac{\partial^2 p}{\partial x^2} + \frac{\partial^2 p}{\partial y^2} + \frac{\partial^2 p}{\partial z^2} = d. \tag{2.46}$$

2.5 Excercises

2.1 Given a velocity variation for the y-component as $v = -ay$, find the corresponding variation of u in the x-direction that satisfies continuity in two dimensions.

2.2 Given the velocities $u = 2xy$, $v = 3yx$ in the x- and y-directions, find the variation of w in the z-direction that satisfies continuity.

2.3 Does the velocity field $u = ay, v = -by$ satisfy continuity in two dimensions?

2.4 Show that integrating the divergence constraint over a rectangular control volume produces Eq. 2.4.

2.5 Consider the momentum balance of inviscid (ideal) flow over a symmetric aerofoil at zero angle of attack in a wind-tunnel (flow is bounded on the sides by the wind-tunnel walls). What are the lift and drag forces on the aerofoil? If the velocity profile at the inlet is uniform, what is the velocity profile at the outlet?

2.6 Consider the same case as Exercise 2.5, but in viscous flow. What are the lift and drag forces on the aerofoil? If the velocity profile at the inlet is uniform, what is the velocity profile at the outlet?

2.7 Consider the momentum balance of fully developed laminar flow in a pipe with circular cross section. If the inlet velocity profile is parabolic, what is the outlet profile? How are pressure gradient and wall shear stress related to each other?

2.8 The pressure drop in a fully developed flow through a pipe with circular cross section of 2-cm diameter is measured as 5 Pa/m. Calculate the wall shear stress and the gradient of the velocity profile at the wall.

[17] Siméon Denis Poisson, French mathematician and physicist, 1781-1840.

2.9 Consider the case of Example 2.7, but in this case flow is uniform at the inlet. Sketch the pressure contours while the flow is developing. Will the wall shear stress be higher or lower compared to the fully developed case?

2.10 Consider a boundary layer that is developing on a flat plate in freestream flow: the velocity will be zero directly on the plate and rapidly increase away from the plate until the freestream velocity is reached. The boundary layer thickness is defined at the distance normal to the plate where the velocity reaches 99% of the freestream velocity. Use the momentum balance to argue whether the boundary layer thickness is constant when the layer is fully developed, or not.

2.11 Derive the differential form of the heat equation (2.43) from the integral form (2.42) using the same approach as taken for the continuity equation (2.11).

3

Discretisation of the equations

In Ch. 2 we have derived the governing equations used in CFD which describe the flow behaviour. We have seen that the formulation of the equations can take two main forms: the integral form considers the sum of fluxes over arbitrary but finite-sized control volumes in the flow field. On the other hand, the differential form relates the derivatives at any point in the flow field, i.e., over infinitesimally small control volumes.

The equations of fluid flow, the Euler or Navier-Stokes equations, are nonlinear second-order partial differential equations. Only for a very small set of cases with very limiting assumptions do we know analytic (i.e., mathematical) solutions to the equations. In all other cases we currently have to rely on approximate solutions. The most successful approach to this approximation is *discretisation* where we approximate the solution over a limited number of points or control volumes.

A simple and straightforward discretisation can be derived from the equations (2.5) and (2.30-2.32), if we limit ourselves to satisfy these equations only at the grid points. The scheme is known as the *finite difference method* and shall serve here as a first example of a discretisation scheme. The analysis of the finite difference method exemplifies very well many basic principles of discretisation that also apply to other discretisation methods.

Most CFD codes are based on the conservative or integral formulation of the equations (2.12) and (2.29) which ensures conservation over the control volume. This is the *finite volume* method. The main advantage of this method is that it is, by construction, conservative. We will introduce the differences between finite volume and finite difference methods in the later parts of this section. All finite volume methods can be expressed as a finite difference method, and it is in that form that we typically analyse the properties of the finite volume method. But since not all finite difference methods are conservative, the conversion from finite difference to a finite volume formulation is not always possible.

In the following we will develop finite difference and finite volume discretisations for the model equations and compare their properties.

3.1 Discretisation of the linear advection equation

As a first equation, let us consider the linear advection equation

$$\frac{\partial u}{\partial t} + a\frac{\partial u}{\partial x} = 0 \tag{2.35}$$

and discretise it using the finite difference method.

3.1.1 Finite difference discretisation of linear advection

If the equation is one-dimensional, a simple choice is to lay out *grid nodes* in an equally spaced row along x. Let us number the nodes with subscript i as $1 \leq i \leq N$. The value of u, also called the *state value* or *state*, at grid node i with coordinate x_i is u_i. The node to the left of i is $i-1$ with u_{i-1}, and the one to the right is $i+1$ with u_{i+1}.

For our simple discretisation with constant distance between nodes, we have $h = x_{i+1} - x_i = x_i - x_{i-1}$. The x-coordinates of the nodes can be written as $x_i = x_0 + ih$, with x_0 being a reference or datum coordinate.

Fig. 3.1 shows how this *discretises* the variation of u along x: we only know the value of u at *discrete* points and need to approximate how u varies between the points. Also, we only have the values of u at the nodes. We do not know the derivatives $\frac{\partial u}{\partial t}$ and $\frac{\partial u}{\partial x}$ which appear in the differential form of the advection equation (2.35). These also need to be approximated.

We could approximate the slope of u at node i as the secant to the curve between two neighbouring mesh points as

$$\frac{\partial u}{\partial x} \approx \frac{u_i - u_{i-1}}{h}; \tag{3.1}$$

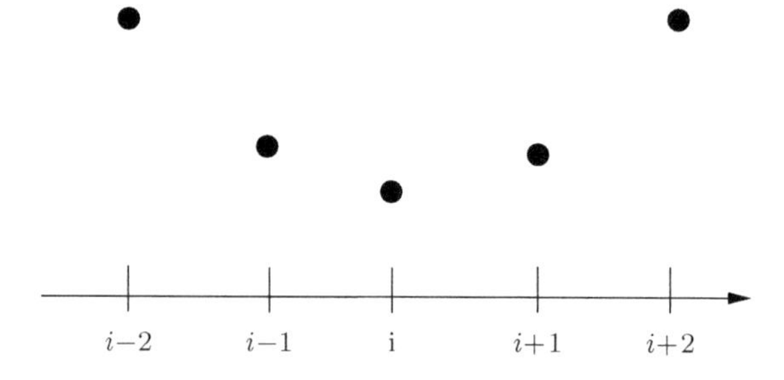

FIGURE 3.1
Discretising a function along a 1-D grid: we only store the values at the nodes $\ldots, i-1, i, i+1, \ldots$, and do not know what happens in between.

such a secant approximation is called a *finite difference* as it is a difference between two points divided by the finite (non-zero, as opposed to infinitesimal) distance between them.[1]

We also need to approximate the time derivative. Let us use a superscript n which indicates the number of the time step called the 'timelevel': $n \geq 0$. Similarly to the space grid let us use a constant time-step $\Delta t = t^{n+1} - t^n$, or $t = t_0 + n\Delta t$. The solution at node i of timelevel n will hence be u_i^n, the one at the following timelevel will be u_i^{n+1}. In the following we will label the timelevel where we know the solution (either from the initially provided solution or from a previous simulation step) as timelevel n, while the unknown solution that we want to compute is at the next timelevel $n+1$.

The time derivative of u at node i can then be approximated as

$$\frac{\partial u}{\partial t} \approx \frac{u_i^{n+1} - u_i^n}{\Delta t}. \tag{3.2}$$

Inserting (3.1) and (3.2) into the advection equation (2.35), and choosing to evaluate the spatial derivative (3.1) at the timelevel n where we know the solution, we find:

$$\begin{aligned} \frac{u_i^{n+1} - u_i^n}{\Delta t} + a\frac{u_i^n - u_{i-1}^n}{h} &= 0 \\ u_i^{n+1} = u_i^n - \frac{a\Delta t}{h}(u_i^n - u_{i-1}^n)&. \end{aligned} \tag{3.3}$$

This method is called the *first-order upwind scheme*: it is first-order accurate in space since we have used the backward difference for the space derivative. We consider the case $a > 0$. The backward difference includes the solution u_{i-1} from node $i-1$ in the update formula: this means we pick the information from upstream or '*upwind*', which is where the information is coming from.[2]

We can simplify and clarify the notation by introducing the *Courant number*[3] $\nu = \frac{a\Delta t}{h}$ named after three pioneers in numerical methods, Courant, Friedrichs and Lewy.[4] We obtain

$$\begin{aligned} u_i^{n+1} &= u_i^n - \nu(u_i^n - u_{i-1}^n) \\ u_i^{n+1} &= (1-\nu)u_i^n + \nu u_{i-1}^n. \end{aligned} \tag{3.4}$$

[1] In Sec. 4.1 we will look in more detail at how this finite difference relates to the exact derivative and consider other approximations.

[2] With numbering from left to right and $a > 0$ the backward difference is upwind. Conversely, if $a < 0$ for that numbering the forward difference is upwind.

[3] The Courant number is often also referred to as the CFL number, since the condition that $\nu \leq 1$ is called the CFL condition.

[4] Richard Courant, German mathematician, 1888-1972, Kurt Otto Friedrichs, German-American mathematician, 1901-1982, Hans Lewy, German-American mathematician, 1904-1988.

3.1.2 Solving the finite difference approximation

How can we solve this equation? The time derivative in the equation describes the evolution of the solution profile $u(x)$ as it is advected with speed a. Clearly this depends on what profile we start with; hence, we need to specify the *initial solution* at timelevel $n=0$. Let us use $u_i^0 = 0$ in all nodes i at $n=0$.

Given the initial solution at $n=0$, we can explicitly solve Eq. 3.3 for the single unknown $u_i^{n+1} = u_i^1$ for each node i at timelevel 1. Having approximated the solution at timelevel 1, we can compute timelevel 2, and so on until we have simulated the desired time interval.

The discretisation we have chosen for the spatial derviative $\frac{\partial u}{\partial x}$ links the nodes i and $i-1$; this also links the first node $i=1$ with $i=0$, which doesn't exist in our grid. With $a > 0$, the flow enters at the left; hence, the value of u at x_0 specifies how the flow enters our area of simulation, the *computational domain*. Specifying the required information at the boundaries of the domain is called a *boundary condition*. In this case, let us assume that the flow enters with $u = 1$ on the left at all times.

Fig. 3.2 shows the evolution of the solution. In this case we advect a block profile made of two steps, the width of the block is discretised with 8 nodes, the numerical simulation is shown with circles, and the exact solution is the shaded line.

We can easily set up the solution of (3.3) in a spreadsheet as shown in Fig. 3.3 for 8 mesh points, focusing on the right flank of the step. The mesh point index i is shown on the top row with $1 \leq i \leq 8$; the column with $i=0$ corresponds to the inflow condition $u = 1$ at the left boundary of the domain. We simulate 12 time-steps with timelevels $1 \leq n \leq 12$ in subsequent rows; the row labelled $n=0$ corresponds to the initial solution. In this case the Courant number has been chosen as $\frac{a\Delta t}{h} = \nu = 0.5$.

The solution in the domain is updated as specified in (3.3). For example, let us compute $u_i^{n+1} = u_2^1$, i.e., the solution in the second mesh point at the first timelevel. The update formula (3.3) involves $u_i^n = u_2^0 = 0$ and $u_{i-1}^n = u_1^0 = 1$. Hence $u_2^1 = (1-\nu)1 + \nu 0 = 0.5$. Pulling the formula across the columns of meshpoints and then down the rows of time-steps automatically translates the required indices in the spreadsheet.

The initial solution has a sharp drop from $u_0^0 = 1$ to $u_1^0 = 0$ over the width of the mesh spacing h. The exact solution for this linear advection equation with the chosen initial and boundary conditions should show that this sharp step is transported through the domain with its shape preserved. As we chose $\nu = 0.5$, it should travel a distance of $\Delta x = a\Delta t = a\nu\frac{h}{a} = \frac{h}{2}$, i.e., half a width of a cell, or half a *mesh width*[5] in each time-step.

The solution for the first time-step $n = 1$ shows that the width of this drop has doubled. It is now spread over 2 mesh widths from $i=0$ to $i=2$. The

[5] A commonly used term for the distance between nodes or the cell width is 'mesh size'. This term is ambiguous, as often it is also used for the number of nodes or cells in a mesh. We hence avoid it here.

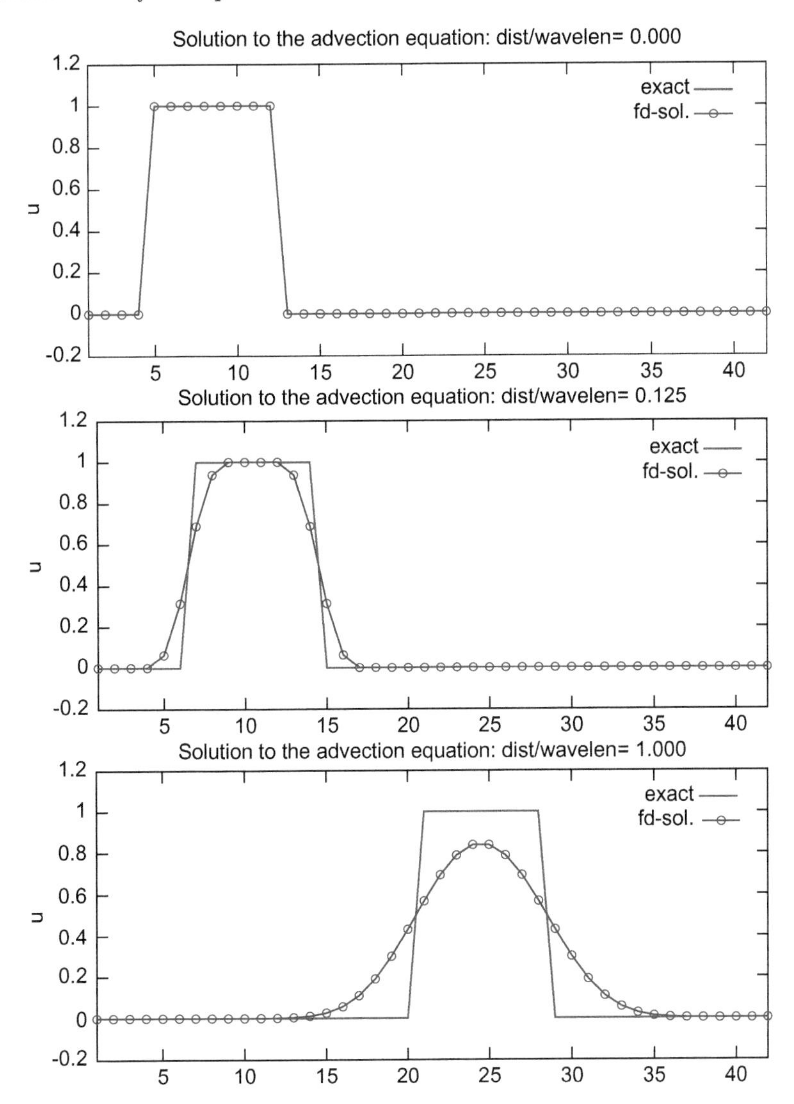

FIGURE 3.2
Advection of a block profile using the first-order upwind scheme. Initial solution (top), after 4 iterations with CFL=0.5 (middle), after advection for 2 block widths (bottom).

sharp profile has been smeared. Further smearing occurs as we proceed with the simulation. At $n=7$ it is smeared over the entire width of the grid with 8 mesh widths. Clearly, the simulation with the first-order upwind method is not very accurate: in deriving the linear advection equation we have neglected all viscous terms from the momentum equation, but we observe effects

	i=0	1	2	3	4	5	6	7	8
n=0	1	0	0	0	0	0	0	0	0
1	1	0.5	0	0	0	0	0	0	0
2	1	0.750	0.250	0	0	0	0	0	0
3	1	0.875	0.5	0.125	0	0	0	0	0
4	1	0.938	0.688	0.313	0.063	0	0	0	0
5	1	0.969	0.813	0.5	0.188	0.031	0	0	0
6	1	0.984	0.891	0.656	0.344	0.109	0.016	0	0
7	1	0.992	0.938	0.773	0.5	0.227	0.063	0.008	0
8	1	0.996	0.965	0.855	0.637	0.363	0.145	0.035	0.004
9	1	0.998	0.980	0.910	0.746	0.5	0.254	0.090	0.020
10	1	0.999	0.989	0.945	0.828	0.623	0.377	0.172	0.055
11	1	1	0.994	0.967	0.887	0.726	0.5	0.274	0.113
12	1	1	0.997	0.981	0.927	0.806	0.613	0.387	0.194

FIGURE 3.3
Finite difference approximation for the linear advection equation with a CFL-number of $\nu = 0.5$.

non-physical or *artificial viscosity* that smear the sharp profile. In the next chapter we will look at alternative discretisations which have less artificial viscosity and should better preserve the solution profile.

We can also see that the initial solution has a minimum value of $u = 0$, and a maximum value of $u = 1$. The exact solution merely transports this profile and so maintains these minimum and maximum values. Indeed, the first-order upwind method does not exceed these bounds; it is *monotonic.* This is an important property, which as we will see is difficult to achieve when discretisations with higher accuracy are used. We will discuss in the next chapter how we can improve both accuracy and monotonicity.

A final observation can be made on the speed of advection. We expect the step to travel half a mesh width in each time-step. Let us use the level of $u = 0.5$ to mark the middle of the wave and track the advection of that level to determine the advection speed. At $n=0$ this level is located halfway between nodes $i=0$ and $i=1$, and at $n=1$ it is found at $i=1$, so it has indeed travelled half a mesh width. Following this over the next time-steps, we can see that the wave travels with the correct speed. The first-order upwind scheme has a zero *dispersion error*; it reproduces the advection speed correctly. Very few other discretisation schemes offer this property, but it is a less important criterion in steady flows or flows without shockwaves.

3.1.3 Mesh refinement

Is there a remedy against the severe smearing due to artificial viscosity that we observed in the spreadsheet solution and its graphical representation in Fig. 3.2? The effect of artificial viscosity is actually less severe on this finer mesh. While the block profile becomes more rounded it is still recognisable. However, the peak value of $u = 1$ is lost by the time the solution is transported for two block widths.

We can refine the mesh further. Fig. 3.4 shows the same case on the mesh from Fig. 3.2, as well as a mesh twice and four times refined. The peak value of the twice refined mesh with 16 nodes per block width nearly captures the peak value correctly; on the mesh with 32 nodes the centre of the block profile is still at $u = 1$.

The number of nodes on each flank of the block from $u = 0$ to $u = 1$ is the same in all cases. As the analysis in Sec. 4.2 will show, the discretisation error of first-order upwind scheme is proportional to the mesh width h, or h^1. The method is *first-order accurate*. Refining the mesh by a factor two will produce half the error while using twice as many points, hence the error per node is the same.

The results in Fig. 3.4 demonstrate that one may have to run very fine meshes with an exceedingly large number of mesh nodes if discretisation errors are to be kept small. For complex flows this can mean that the computer's memory may not be sufficient to achieve the desired accuracy. Sec. 4.3 will consider alternative discretisations to the first-order upwind method where the error reduces faster with mesh refinement. CFD typically uses *second-order accurate* discretisations where the discretisation error is proportional to h^2, hence reduced by a factor of 4 when refining the mesh by a factor of 2.

3.1.4 Finite volume discretisation of the 1-D advection

The derivation of the finite-difference discretisation of the advection equation in Sec. 3.1.1 started from the strong or differential form of the equation (2.35). However, while convenient to analyse and easy to follow, the finite difference method is not conservative as discussed in more detail in Sec. 3.2. The finite volume method is by far the most popular discretisation approach in CFD because it is conservative by construction.

Let us use for the finite volume method the same grid as used for the finite difference method, but define the control volume of node i as the 'volume'[6] between $\frac{1}{2}(x_{i-1}+x_i) = x_i - \frac{h}{2} = x_{i-\frac{1}{2}}$ on the left and $\frac{1}{2}(x_i+x_{i+1}) = x_i + \frac{h}{2} = x_{i+\frac{1}{2}}$ on the right as shown in Fig 3.5. The advection equation balances the net flow in or out of the volume over time with how much is accumulated in the volume. In the steady state there is no change in time. In this case it is sufficient to consider the spatial variation only. However we will need to consider the transient to the steady state, so the control volume also has an extension in time which we will take as ranging from $t = t^n$, typically a timelevel where

[6]Even though we are in one dimension, let us keep the notion of 'volume' for the intgration over the control volume, even though in 1-D this is only a line.

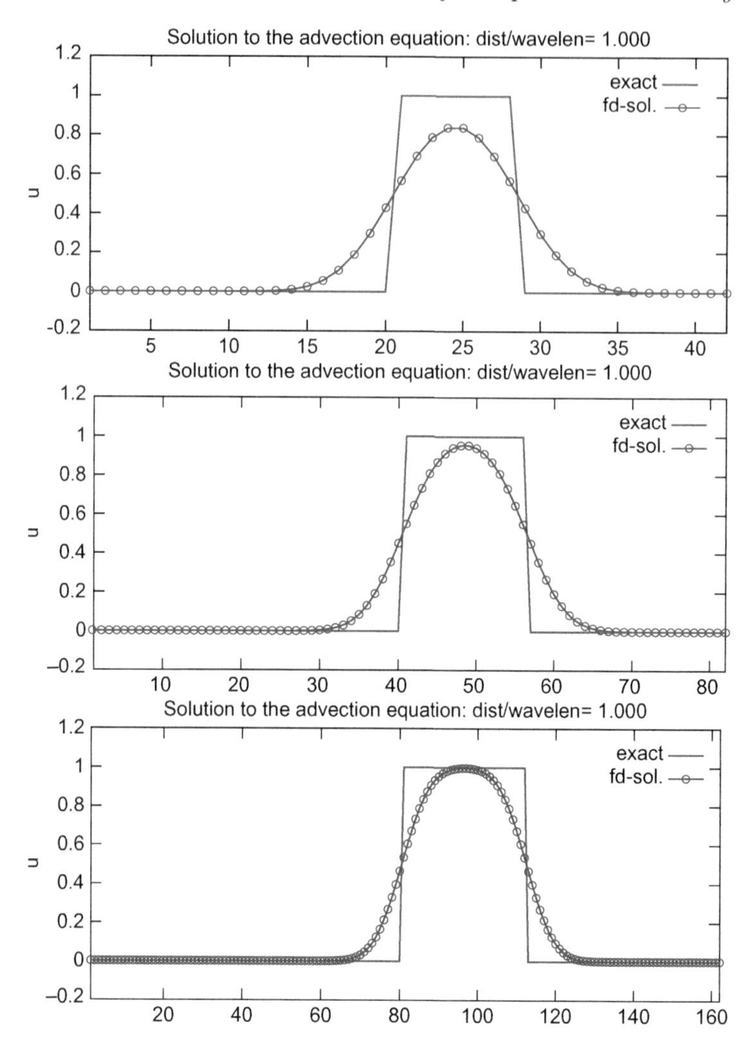

FIGURE 3.4
Advection of a block profile with the first-order upwind scheme for two block widths with CFL=0.5. Mesh with 8 (top), 16 (middle) and 32 nodes per block (bottom).

we know the solution, to $t = t^n + \Delta t = t^{n+1}$ which is the timelevel where the solution is to be computed.

Integrating the flux balance (2.36) over the control volume we have

$$\int_{t^n}^{t^{n+1}} \int_{x-\frac{1}{2}}^{x+\frac{1}{2}} \left(\frac{\partial u}{\partial t} + \frac{\partial f}{\partial x} \right) dx dt = 0. \tag{3.5}$$

As a first simplification, let us assume that over the time interval Δt the

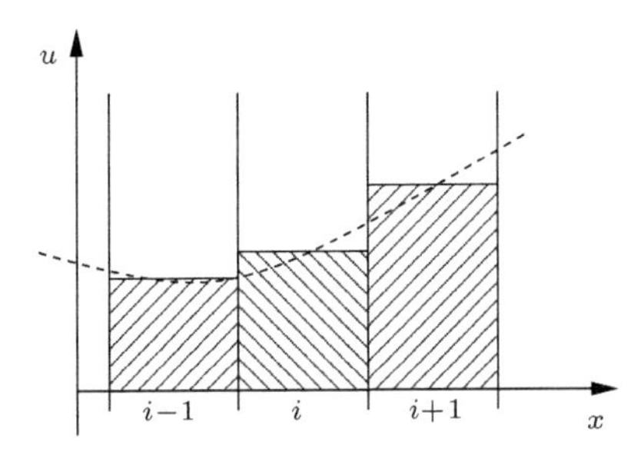

FIGURE 3.5
Control volume for a finite-volume discretisation of the advection equation.

fluxes f and the rate of change $\partial u/\partial t$ are constant in time. Carrying out the integration in time and dividing by the time-step Δt one finds

$$\int_{x-\frac{1}{2}}^{x+\frac{1}{2}} \left(\frac{\partial u}{\partial t} + \frac{\partial f}{\partial x} \right) dx = 0. \tag{3.6}$$

If we take the state u to be represented as a constant in each control volume, then the flux f can be considered constant in the short time interval while the constant solution has not been transported out of the volume. Then

$$\int_{x-\frac{1}{2}}^{x+\frac{1}{2}} \frac{\partial u}{\partial t} dx + f(x_{i+\frac{1}{2}}) - f(x_{i-\frac{1}{2}}) = 0. \tag{3.7}$$

We have so far not made any assumptions how the property u varies over the control volume. The simplest choice is that it is constant within the control volume, with a jump at the interfaces $x+\frac{1}{2}$ to the next volume (cf. Fig 3.5). Fig. 3.6 shows how such a *piecewise constant* data representation can reproduce spatial variations. In this geological formation volcanic basalt rock is crystallised into vertical hexagonal columns which erode in horizontal layers, leaving the tops of each column flat. The jumps between the hexagonal cells can clearly be observed.

We are interested in the time evolution of the average in the cell; hence, we can assume that the time derivative is constant over the control volume and the advection equation becomes

$$\begin{aligned}
\int_{x_{i-1/2}}^{x_{i+1/2}} dx \frac{\partial u}{\partial t} + f(x_{i+\frac{1}{2}}) - f(x_{i-\frac{1}{2}}) &= 0 \\
h\frac{\partial u}{\partial t} + f_{i+1/2} - f_{i-1/2} &= 0, \\
\frac{\partial u}{\partial t} + \frac{f_{i+1/2} - f_{i-1/2}}{h} &= 0.
\end{aligned} \tag{3.8}$$

FIGURE 3.6
Giant's Causeway: piecewise constant states can approximate a spatial variation.

Separating the spatial terms from the time derivative in Eq. 3.8

$$\frac{\partial u}{\partial t} = \frac{f_{i-1/2} - f_{i+1/2}}{h} \tag{3.9}$$

illustrates the conservation principle very clearly: the net balance of fluxes is equal to the rate of change, the *residual.* [7] The residual in finite volume methods is not a measure of error, it is a measure of unsteadiness. In a simulation for the steady state it should be near zero, but would typically be non-zero for unsteady flows.

Fluxes need to be computed at left and right interfaces of the control volume $i-\frac{1}{2}, i+\frac{1}{2}$. The flux of u in the linear advection equation is $f = au$ with $a = \text{const}$, but u is discontinuous across the interface as in general $u_{i-1} \neq u_i$.

A sensible choice to compute the flux f is to take the information from where it is coming from, from 'upwind'.[8] In our example with $a > 0$ infor-

[7] Residual does take many meanings in numerical methods. In finite volume method it typically refers to the net flux balance which is equal to the rate of change.

[8] Other choices will be discussed in Sec. 4.3.1.

mation comes from the left. Hence a good choice is $f_{i-\frac{1}{2}} = au_{i-1} = f_{i-1}$ and $f_{i+\frac{1}{2}} = au_i = f_i$. Using the same forward difference approximation for the time derivative as for the finite difference scheme (3.3) the finite volume discretisation becomes

$$\frac{u_i^{n+1} - u_{i-1}^n}{\Delta t} + \frac{1}{h}(f_{i-1}^n - f_i^n) = 0$$

$$u_i^{n+1} = u_{i-1}^n - \frac{\Delta t}{h}(f_i^n - f_{i-1}^n). \tag{3.10}$$

Note that here we have also chosen to evaluate the fluxes at the known timelevel n, so we can solve explicitly for the unkonwn u_i^{n+1}.

3.1.5 Solving the finite volume approximation

The approach to solving the equations of the finite volume method is very similar to the finite difference formulation (3.3): we need an initial solution at $n = 0$ and a boundary condition at $i = 0$. Given that, we can compute the single unknown in the equation u_i^{n+1} explicitly for each node i at the new timelevel.

Similarly to the finite difference approximation we can set up the finite volume method in a spreadsheet as shown in Fig. 3.7. As in the finite difference example, we find the grid nodes $1 \leq i \leq 8$ running in rows across, with the index i in the top row. The interfaces between control volumes $i \pm \frac{1}{2}$ are listed between the mesh points as 1.5, 2.5, etc. The node with $i=0$ in grey corresponds to the inflow condition. Timelevels run down in rows, with the index n in the first column; $n=0$ in grey is the initial condition. Each timelevel comprises two rows, the row with the state u^n followed by a row with the interface fluxes for that timelevel in pale-shaded cells.

The finite volume method is guaranteed to be conservative if we ensure that the flux at $f_{i+\frac{1}{2}}$ exiting volume i is the same as the flux entering volume $i+1$. If this is ensured, the flux between volumes cancels exactly, the flux that leaves the volume to the left of the interface enters the volume to the right. In a flux balance over the entire domain all interior fluxes cancel, leaving only contributions from the boundaries. As an analogy one could think of the barrels of a telescope that slide into each other. Pushed together only the outermost barrel is visible; hence, the fluxes are said to *telescope.*

3.1.6 Finite difference vs. finite volume formulations

It is instructive to compare the finite volume discretisation (3.10) with the finite difference discretisation (3.3). In the finite difference form we have directly approximated the derivatives and assemble the terms in the strong form of the differential equation. We have seen in Secs. 2.2.2 and 2.3 that the strong form of the equation is derived by assembling the flux balance over a control

	i=0	*0.5*	1	*1.5*	2	*2.5*	3	*3.5*	4	*4.5*	5	*5.5*	6	*6.5*	7	*7.5*	8	*8.5*
n=0	1		0		0		0		0		0		0		0		0	
flux		*1*		*0*		*0*		*0*		*0*		*0*		*0*		*0*		*0*
n=1	1		0.5		0		0		0		0		0		0		0	
		1		*0.5*		*0*		*0*		*0*		*0*		*0*		*0*		*0*
2	1		0.75		0.25		0		0		0		0		0		0	
		1		*0.75*		*0.25*		*0*		*0*		*0*		*0*		*0*		*0*
3	1		0.88		0.5		0.13		0		0		0		0		0	
		1		*0.88*		*0.50*		*0.13*		*0*		*0*		*0*		*0*		*0*
4	1		0.94		0.69		0.31		0.06		0		0		0		0	
		1		*0.94*		*0.69*		*0.31*		*0.06*		*0*		*0*		*0*		*0*
5	1		1		0.81		0.5		0.19		0.03		0		0		0	
		1		*1*		*0.81*		*0.50*		*0.19*		*0.03*		*0*		*0*		*0*
6	1		1		0.89		0.66		0.34		0.11		0.02		0		0	
		1		*1*		*0.89*		*0.66*		*0.34*		*0.11*		*0.02*		*0*		*0*
7	1		1		0.94		0.77		0.5		0.23		0.06		0.01		0	
		1		*1*		*0.94*		*0.77*		*0.5*		*0.23*		*0.06*		*0.01*		*0*
8	1		1		1		0.86		0.64		0.36		0.14		0.04		0	
		1		*1*		*1*		*0.86*		*0.64*		*0.36*		*0.14*		*0.04*		*0*
9	1		1		1		0.91		0.75		0.5		0.25		0.09		0.02	
		1		*1*		*1*		*0.91*		*0.75*		*0.50*		*0.25*		*0.09*		*0.02*
10	1		1		1		0.95		0.83		0.62		0.38		0.17		0.05	
		1		*1*		*1*		*0.95*		*0.83*		*0.62*		*0.38*		*0.17*		*0.05*
11	1		1		1		1		0.89		0.73		0.5		0.27		0.11	
		1		*1*		*1*		*1*		*0.89*		*0.73*		*0.50*		*0.27*		*0.11*
12	1		1		1		1		0.93		0.81		0.61		0.39		0.19	

FIGURE 3.7
Finite volume approximation for the linear advection equation with $\nu = 0.5$.

volume and then shrinking its size to zero. But on a mesh of discrete points we are unable to take this limit to zero. We can at most shrink the size to the distance between mesh points, the mesh width h. Whenever there are strong variations in the solution we must expect significant errors in the finite difference formulation.

In the finite volume formulation we directly discretise the sum of fluxes of a conserved quantity; in the case of the advection equation the scalar u is conserved. Unlike the strong form of the equations (2.35), the weak form (2.36) exactly conserves the transported property. There is no need to shrink the control volume to zero to obtain conservation.

We have however made an assumption on how the state u varies within the control volume: in this example it was taken to be constant over the volume. This approximation has in turn been used to approximate the fluxes. Since the state u is discontinuous at the interface there is also no unique value for the flux. By choosing a flux at the interface to have a value somewhere in between the fluxes of the two constant states either side, we have made a further approximation. Hence while conservation of the finite volume method is exact, the approximation of the fluxes still carries significant discretisation errors. But most importantly, the method explicitly allows and can handle fields that are not smooth but have jumps, so it is very suitable for equations that allow discontinuities such as the Euler and Navier-Stokes equations.

The finite difference method, on the other hand, is not necessarily conservative since it does not use fluxes and hence does not always ensure that what leaves one node passes to its neighbour. Rapid changes in the solution could lead to a creation or destruction of a conserved quantity such as mass, momentum or energy. Conservation is not a crucial property in smooth flows (or, e.g., in structures analysis), but is essential when strong variations are present such as e.g., in boundary layers, shocks or flame fronts.

In the case of linear advection we can actually factor the constant advection speed out of the flux and write $\frac{\partial f}{\partial x} = a\frac{\partial u}{\partial x}$, $a = \text{const}$. There are no variations in the transport velocity. Inserting this into the finite volume discretisation (3.10) we find that for constant advection speed a the finite difference and the finite volume formulations are actually identical. Comparing the two spreadsheet solutions for the finite difference method in Fig. 3.3 and for the finite volume method in Fig. 3.7 confirms that the results are identical.

This will not be true in general. In the momentum equation (2.33) the advection term is $u\frac{\partial u}{\partial x}$; the advective part of the flux is $f = \frac{1}{2}u^2$. The advection term is non-linear and we cannot factor out the advection speed u as it is not constant. In this more general case the finite volume method is still guaranteed to be conservative as the control volume formulation is exact, but the finite difference formulation is not necessarily conservative, as will be shown for Burgers' equation in Sec. 3.2 and the errors in the approximation can result in an increase or loss of the conserved quantity.

In general, we can always *linearise* the flux derivative $\frac{\partial f}{\partial x} = \frac{\partial f}{\partial u}\frac{\partial u}{\partial x} = a(u)\frac{\partial u}{\partial x}$ and turn any finite volume method into a finite difference method. This is actually how we analyse the characteristics of a finite volume method. On the other hand, we cannot turn any finite difference method into a finite volume method, which reflects the fact that in general finite difference methods are not conservative. In summary we can say

- the finite difference method approximates derivatives in the strong form of the equations,
- the finite volume discretisation considers the flux balance based on the conservative form,
- the finite volume method is equivalent to applying control-volume theory on all the volumes,
- both forms are identical for linear equations as with $a = const$,
- but they will differ for non-linear equations such as the momentum equation,
- the finite volume method is conservative by construction: between neighbouring volumes: flux in = flux out.

3.2 Burgers' equation: non-linear advection and conservation

In the preceding sections we have seen that for the linear advection equation the discretisations with the finite difference and the finite volume method were identical because the advection speed a is constant. Burgers' equation has a non-linear convection term similar to the momentum equation. Sec. 3.2 introduced its differential form:

$$\frac{\partial u}{\partial t} + u\frac{\partial u}{\partial x} = 0, \tag{2.39}$$

and its conservative form:

$$\frac{\partial u}{\partial t} + \frac{\partial f}{\partial x} = 0 \quad \text{with} \quad f = \frac{1}{2}u^2. \tag{2.40}$$

Let us first develop a finite difference approximation using (2.39), borrowing principally from the finite difference discretisation of the linear advection eq. (3.3). To adapt to Burgers' equation, we replace the constant advection speed a with the local speed u_i when differencing for node i,

$$\begin{aligned} \frac{u_i^{n+1} - u_i^n}{\Delta t} + u_i \frac{u_i^n - u_{i-1}^n}{h} &= 0 \\ u_i^{n+1} = u_i^n - \frac{u_i \Delta t}{h}(u_i^n - u_{i-1}^n), \end{aligned} \tag{3.11}$$

and similarly for node $i+1$,

$$u_{i+1}^{n+1} = u_{i+1}^n - \frac{u_{i+1}\Delta t}{h}(u_{i+1}^n - u_i^n). \tag{3.12}$$

The finite volume discretisation of Burgers' equation can follow (3.8), using our standard forward difference in time (3.2)

$$u_i^{n+1} = u_i^n - \frac{\Delta t}{h}\left(f_{i+1/2} - f_{i-1/2}\right) = 0, \tag{3.13}$$

and similarly for node $i+1$,

$$u_{i+1}^{n+1} = u_{i+1}^n - \frac{\Delta t}{h}\left(f_{i+3/2} - f_{i+\frac{1}{2}}\right) = 0. \tag{3.14}$$

The flux $f_{i+\frac{1}{2}}$ appears in both 3.13 and 3.14. The finite volume method does not specify how to discretisatise this flux, but it does insist that the flux $f_{i+\frac{1}{2}}$ of cells i is the same as flux $f_{i-\frac{1}{2}}$ for cell $i+1$.

Comparing the finite volume updates for node i in (3.13) to the finite difference update for i in (3.11) we find that the flux $f_{i+\frac{1}{2}}$ for node i in the finite difference method is actually $f_{i+\frac{1}{2}} = u_i u_i$,[9] while the flux $f_{i-\frac{1}{2}}$ for cell $i-1$ is $f_{i-\frac{1}{2}} = u_{i+1} u_i$. In the general case of $u_i \neq u_{i+1}$ the fluxes are not identical, and the finite difference discretisation (3.11) is not conservative.

Let us change the way we approximate the advection speed in the finite difference approximation: instead of using the nodal value u_i at node i as done in (3.11), we could choose to also use an upwind approximation for that speed:

$$u_i^{n+1} = u_i^n - \frac{\Delta t}{h}(u_i^n u_i^n - u_{i-1}^n u_{i-1}^n). \tag{3.15}$$

This seems arbitrary at first, but analysing the resulting fluxes as for (3.11), one can see that (3.15) is conservative. It actually is equivalent to a finite volume formulation with a flux of $f_{i+\frac{1}{2}} = u_i u_i$.

3.3 Heat equation in 1-D

This chapter demonstrates the discretisation of second derivatives using the example of the heat equation (2.45).

3.3.1 Discretising second derivatives

We need to add the discretisation of a second derivative to our taxonomy of discretisations. Assume that we know the temperatures T at the points i. We could then compute a first derivative T_x at the interfaces $i - \frac{1}{2}$ and $i + \frac{1}{2}$:

$$T_x|_{i-\frac{1}{2}} = \frac{T_i - T_{i-1}}{h}, \quad T_x|_{i+\frac{1}{2}} = \frac{T_{i+1} - T_i}{h},$$

and then take the derivative again of these mid-point first derivatives to obtain a second derivative T_{xx}:

$$T_{xx}|_i = \frac{T_x|_{i+\frac{1}{2}} - T_x|_{i-\frac{1}{2}}}{h} = \frac{T_{i-1} - 2T_i + T_{i+1}}{h^2}. \tag{3.16}$$

Of course there are other choices, e.g., one could compute the first derivatives not at the interfaces $i \pm \frac{1}{2}$, but at the neighbouring points $i \pm 1$, which would include the values of T at $i \pm 2$. These first derivatives could then be used to compute the second derivative at i. However, this discretisation results in a stencil (molecule) of five points rather than the three points of (3.16). In fact, (3.16) is the most compact discretisation for the second derivative in one dimension, i.e., its stencil has the smallest width.

[9] For ease of comparison, the factor $\frac{1}{2}$ is subsumed into the time-step Δt.

3.3.2 1-D Heat equation, differential form

As an example, consider the steady heat equation in differential form, (2.44), in one dimension,

$$\frac{\partial^2 T}{\partial x^2} = 0, \tag{2.44}$$

simulating heat conduction along a rod.

Let us consider a case with unit length $0 \leq x \leq 1$ where the ends of the rod are kept at fixed temperatures, $T(x=0)=0, T(x=1)=1$. The solution we expect is a linear temperature variation between the left and right ends.

For the spatial derivative in (2.44) using (3.16) we find for the second derivative in x at point i

$$T_{xx}|_i = \frac{T_{i-1} - 2T_i + T_{i+1}}{h^2}.$$

We also need to assign boundary conditions to reflect the fact that the temperature is fixed at the ends. In this simple case the values at the end of the rod are given, so we can directly use them in our discretisation,

$$T_0 = T(0) = 0, \qquad T_{N+1} = T(1) = 1. \tag{3.17}$$

3.3.3 Solving the 1-D heat equation

The heat equation can be solved exactly, but we seek here a numerical approximation. Similar to what is typically used for the flow equations, we consider the unsteady form of the heat equation,

$$\frac{\partial T}{\partial t} + \frac{k}{c_p \rho}\frac{\partial^2 T}{\partial x^2} = 0 \tag{2.43}$$

and observe the evolution of the solution for a time that is long enough to reach a sufficient approximation to the steady state.

Discretising time with a forward difference similar to what was used in Sec. 3.1.2,

$$\frac{\partial T}{\partial t} \approx \frac{T_i^{n+1} - T_i^n}{\Delta t}, \tag{3.2}$$

and evaluating the second derivatives at the known timelevel n, we obtain an explicit discretisation,

$$T_{i,j}^{n+1} = T_{i,j}^n - \frac{\Delta t\, k}{c_p \rho h^2}\left(T_{i-1}^n - 2Tn_i + T_{i+1}^n\right)$$

or

$$T_{i,j}^{n+1} = cT_{i-1}^n + (1-2c)T_i^n + cT_{i+1,j}^n \tag{3.18}$$

using the shorthand $c = \frac{\Delta t\, k}{c_p \rho h^2}$. Similar to what was discussed in Sec. 3.1.2, we find that for $c \leq 1/4$, all coefficients in the finite difference formula are positive;

	i=0	1	2	3	4	5	6	7	8	9
n=0	0	0	0	0	0	0	0	0	0	1
1	0	0	0	0	0	0	0	0	0.5	1
2	0	0	0	0	0	0	0	0.25	0.5	1
3	0	0	0	0	0	0	0.125	0.25	0.625	1
4	0	0	0	0	0	0.063	0.125	0.375	0.625	1
5	0	0	0	0	0.031	0.063	0.219	0.375	0.688	1
6	0	0	0	0.016	0.031	0.125	0.219	0.453	0.688	1
7	0	0	0.008	0.016	0.07	0.125	0.289	0.453	0.727	1
8	0	0.004	0.008	0.039	0.07	0.18	0.289	0.508	0.727	1
9	0	0.004	0.021	0.039	0.109	0.18	0.344	0.508	0.754	1
10	0	0.011	0.021	0.065	0.109	0.227	0.344	0.549	0.754	1
11	0	0.011	0.038	0.065	0.146	0.227	0.388	0.549	0.774	1
12	0	0.019	0.038	0.092	0.146	0.267	0.388	0.581	0.774	1
102	0	0.111	0.221	0.332	0.443	0.554	0.666	0.777	0.888	1
103	0	0.111	0.222	0.332	0.443	0.554	0.666	0.777	0.889	1

FIGURE 3.8
Solution to the 1-dimensional heat equation on 8 points.

in this case (3.18) is an averaging formula which guarantees boundedness, and the solution is monotonic. The link between the size of the time-step and stability is discussed in more detail in Sec. 4.3.

We can program this in a spreadsheet, similar to what was done for the 1-D advection equation in Sec. 3.1.2. Results for $N = 8$ internal points are shown in Fig. 3.8. We can observe that it takes over 100 iterations to reach a state where the values no longer change to three digits. Better iterative schemes that converge in fewer iterations than the explicit time-stepping used here are discussed in Sec 4.5.

3.4 Advection equation in 2-D

This section presents the discretisation in more than one dimension using as an example the linear advection equation introduced in Sec. 2.4.1. We shall first consider a simple regular structured grid of rectangular cells and then discuss how to extend the model to the general case of an arbitrary unstructured grid such as e.g., formed of triangles. Sec. 7.4 discusses the advantages of unstructured grids in detail.

The domain is the unit square $0 \leq x, y \leq 1$ with N points both in x and y, as shown in Fig. 3.9.

Let us use the boundary conditions $u(0, y) = 0$ on the left, $u(0 \leq x \leq 0.5, 0) = 0$, as well as along the left and right quarters of the bottom and

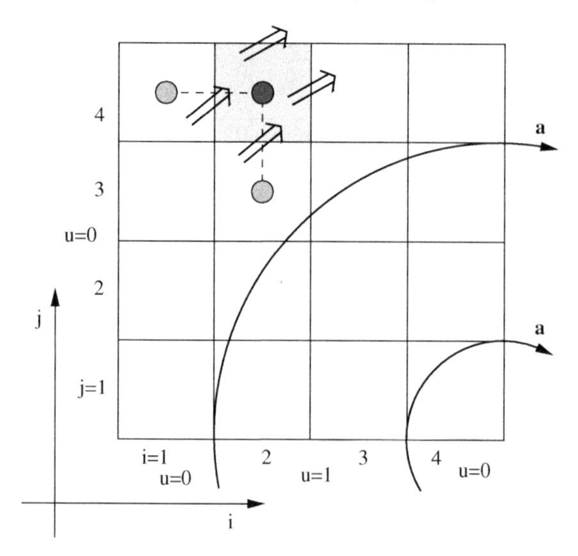

FIGURE 3.9
Discretisation of the advection equation for a passive scalar u with the advection speed **a** on a unit square of $N = 4$ points. The values of u are located at the centres of the cells.

$u(x < 0.25, 0.75 < x) = 1$. Along the centre of the bottom face let us impose inflow with $u(0.25 \leq x \leq 0.75, 0) = 1$. Hence the exact solution in the steady state is $u = 0$ for $r > 0.75, r < 0.25$ and $u = 1$ for $0.25 \leq r \leq 0.75$ with r being the distance from the rotation origin at $x = 1, y = 0$.

3.4.1 Discretisation on a structured grid

We can divide this simple geometry into a regular grid of squares, indexing the points in the x-directions with i, in the y-direction with j; hence $T_{1,1}$ corresponds to the temperature in the bottom left corner. Owing to its regular structure where indices of points can be computed based on the coordinates, this is called a *structured grid.* Our discretisation assumes that the values $u_{i,j}$ are located at the centers of each of the cells of the grid. This *cell-centred* approach is typically chosen for incompressible flow solvers.[10] Let us consider the case of $a = (y, 1 - x)$; hence, the information is travelling on circular arcs around the point $x = 1, y = 0$ from the bottom left to the top right as shown by the arrows for **a** in Fig. 3.9.

We have seen in Sec. 3.1.4 that the finite difference method is not necessarily conservative for non-constant **a** while the finite volume discretisation

[10] We could also locate the unknowns at the nodes of the grid, the node-centred approach often used for compressible flow solvers. The pros and cons of these two approaches are discussed e.g., in [4], but will not be discussed here.

guarantees conservation, and hence we use the finite volume method here. Let us recall the integral form of the advection equation (2.38) that the finite volume method is based on:

$$\iint_V \frac{\partial u}{\partial t} dV + \oint \mathbf{a}u \cdot \mathbf{n}\, dS = 0. \tag{2.38}$$

If we assume that u is constant inside each cell, the piecewise constant data respresentation discussed in Sec. 3.1.4, the integral of the rate of change becomes

$$\iint_V \frac{\partial u}{\partial t} dV = \frac{\partial u}{\partial t} \iint_V dV = A\frac{\partial u}{\partial t} \tag{3.19}$$

where A is the volume (or here the area[11]) of the cell. Using the forward difference in time on this square grid of mesh width h we obtain

$$A\frac{\partial u}{\partial t} = A\frac{u_{i,j}^{n+1} - u_{i,j}^{n}}{\Delta t}.$$

In our structured grid every cell has four straight faces of length h, so the surface integral over the fluxes, also called the *residual* can be computed as

$$\oiint \mathbf{a}u \cdot \mathbf{n}\, dS = \sum_k \mathbf{a}_k u_k \cdot \mathbf{n}_k S_k = \sum_k \mathbf{f}_k \cdot \mathbf{n}_k S_k$$

where the subscript $1 \leq k \leq 4$ refers to each of the four faces of the cell, $f_k = \mathbf{a}_k u_k$ is the flux, $\mathbf{n}_k$ is the outward-oriented face normal and S_k is the area of the face. The four fluxes that contribute to the residual in cell $i = 2, j = 4$ are shown in Fig. 3.9.

Let us choose the first-order upwind scheme as used in the one-dimensional example in Sec. 3.1.4; hence the flux at the interface between the cells $\mathbf{f}$ will be based on the state in the cell that is upwind of the interface, but will be the same for both cells that share the interface. Fig. 3.9 shows the upwind stencil for the point $i = 2, j = 4$. The flux at the western interface $\mathbf{f}_{1+\frac{1}{2},4}$ is computed using the state $u_{1,4}$,

$$\mathbf{f}_{1+\frac{1}{2},4} = \mathbf{a}_{1,4} u_{1,4}$$

while the flux at the southern interface $\mathbf{f}_{2,3+\frac{1}{2}}$ is based on $u_{2,3}$,

$$\mathbf{f}_{2,3+\frac{1}{2}} = \mathbf{a}_{2,3} u_{2,3}.$$

Similarly, the fluxes at the eastern and northern interfaces of cell 2,4 are based

[11] Most of CFD happens in 3-D, where fluxes cross a face with an area to enter a control volume. To avoid confusion, let us adopt the same notation in 2-D.

on $u_{2,4}$. For an interior point on this grid the first-order upwind scheme then becomes

$$\iiint_V \frac{\partial u}{\partial t} dV + \oiint \mathbf{a}u \cdot \mathbf{n}\, dS = 0$$

$$A_{i,j} \frac{u_{i,j}^{n+1} - u_{i,j}^{n}}{\Delta t} + \sum_{k=1}^{4} \mathbf{f}_k \cdot \mathbf{n}_k S_k$$

$$u_{i,j}^{n+1} = u_{i,j}^{n} - \frac{\Delta t}{A_{i,j}} \sum_{k=1}^{4} \mathbf{f}_k \cdot \mathbf{n}_k S_k. \tag{3.20}$$

Boundary conditions at the left $x = 0$ can be imposed through the flux $\mathbf{f} = \mathbf{a}u = 0$ at the western faces $\frac{1}{2}, j$ of the leftmost cells, $1, j$. Similarly at the southern faces where $u = 0$, i.e., $0.25 < x, x0.75 < x$, the flux can be imposed as $\mathbf{f} = \mathbf{a}u = 0$. The flux at the bottom central part where $u = 1$ is non-zero: $\mathbf{f} = \mathbf{a}u = (y, 1 - x)^T$ for $y = 0, 0.25 \leq x \leq 0.75$.

The top and right boundaries in Fig. 3.9 are outflow boundaries, so the 'upwinded' flux depends on the interior point. We can compute these fluxes in the same ways as the interior ones.

The Appendix 10.1 provides the Matlab™ source code for this implementation:

- In the finite volume method the flux across each face between cells is the same for both side. It is this that ensures conservation. We can save computational effort by looping over all faces and computing the flux only once, adding the flux to one cell, subtracting from the other.

- Boundary fluxes can be computed by looping over all boundary faces and imposing either the known flux at inflow boundaries, or computing the exiting flux from the interior cell at outflow boundaries.

- When all faces have been visited, the flux balance or residual has been calculated. This residual is equivalent to the average rate of change in the cell (see (2.38)), and we can use it to compute the update for the state using (3.20).

- We are looking for the steady state solution in this example, but running the code we can observe that the residuals will never become fully zero. The floating point calculations on the computer have a finite precision. Typically for double-precision arithmetic as used by Matlab this will be 14-15 digits, so once the residuals have reached around 10^{-15} round-off error will introduce random fluctuations and the residuals will not converge to smaller values.

  However, when considering engineering accuracy of three or four digits, the values of the solution will have stopped to change at any relevant scale much earlier and we can reduce computational effort by stopping the iterations at that stage. Choosing an appropriate convergence level requires knowledge of the particular case. This is further covered in Sec. 4.5. For the results presented here a convergence threshold of 10^{-7} for the root mean square (RMS) of the residuals was used.

Fig. 3.10 shows the expected significant smearing of the initially sharp profile on meshes of 10x10, 20x20 and 40x40 cells. Fig. 3.11 demonstrates how mesh refinement improves the solution; only the solution on the grid with 40x40 cells manages to reach the peak value of $u = 1$ in the exit plane. However, while in our 1-D advection examples in Sec. 3.1.3 reducing the mesh width by half meant twice as many cells to compute and store, in the 2-D case

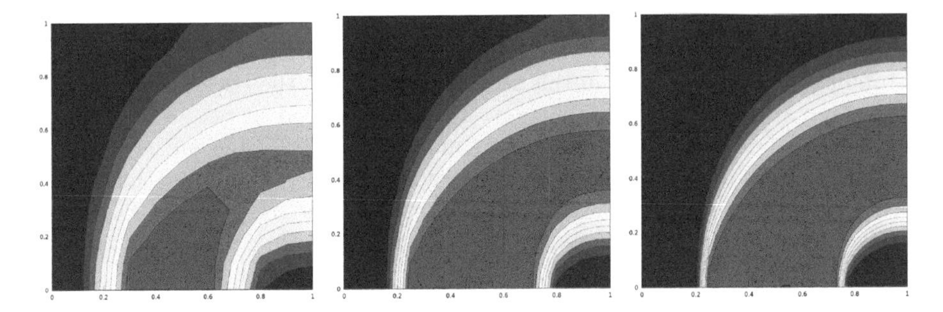

FIGURE 3.10
Circular advection of a block profile using a first-order upwind finite volume scheme. 10x10 mesh (left), 20x20 (middle), 40x40 (right).

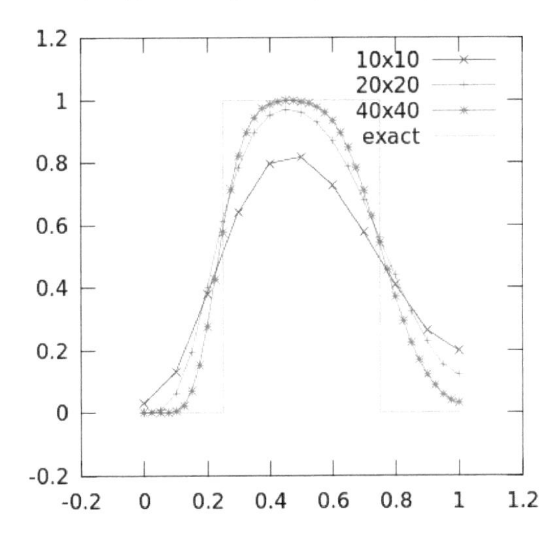

FIGURE 3.11
Exit profiles at $x = 1$ for circular advection of a block profile.

we have to invest 4 times as many cells to reduce h by half. In 3-D it would be a factor of 8. Clearly, improving the solution by uniform mesh refinement will quickly become prohibitively expensive in realistic 3-D cases. Ch. 8 will discuss in more detail how to assess and control errors in the approximation.

3.5 Solving the Navier-Stokes equations

In the preceding sections of this chapter we have seen how to devise a finite volume method for the one- and two-dimensional linear advection equations. We can now understand in principle how the finite volume method would be used for the Navier-Stokes equations in 2-D and 3-D in one of the commercial or open-source solvers which are most often used in CFD. The basic understanding brought together in this section will provide the background for the reader to run first CFD simulations. This is expanded in Sec. 3.6.1 toward arbitrary grids which have become standard in industrial application, and discusses how to use pseudo-time-stepping techniques to compute the steady state. The final sections in this chapter discuss how the system of Navier-Stokes equations is solved for compressible and for incompressible flows. The remaining chapters of the book look at the material in more detail.

3.6 The main steps in the finite volume method

Let us begin with bringing together the main steps of the finite volume method. This basic overview should enable the reader to understand in principle the steps when working through a tutorial for a CFD solver for running a case. As the reader's experience with the software progresses, questions on the background will arise, which the subsequent chapters will address.

- The area covered by the flow simulation, the *domain*, is "tiled" into small finite volumes or *mesh cells* which don't overlap. We refer to the collection of these cells and the associated boundary faces as the *mesh*. The image on the front cover of a grid around an aerofoil is an example for such a mesh. The conserved quantities such as mass, momentum and energy are conserved as they are transported across the cells of the mesh.

- An *initial solution* has to be set, as the equations describe the evolution in time of the flow. Tracing this evolution requires a starting point.

- *Fluxes* of the conserved quantities are calculated between neighbouring cells. Fluxes also need to be computed at the boundaries of the domain to provide the required information for flow entering or leaving the domain. In the finite volume method the flux leaving one cell either enters the neighbouring cell or passes through a boundary. Boundary conditions are discussed in Ch. 5. This is the key advantage of the finite volume method: by construction it is guaranteed to be conservative.

These fluxes depend on the states u in the two neighbouring cells to the left and right of the flux interface.[12]

- The next step is to compute for each finite volume or cell the sum of the fluxes entering or exiting over all its sides, as well as the integral of acting forces. This *residual* is equal to the average rate of change in the volume, e.g., if there is more mass entering than exiting, the net mass in the cell increases over time. If the sum of fluxes is zero, the cell is in balance and the state won't change: what enters the volume is equal to what exits.

- To advance the solution in time the average state in each finite volume needs to be updated frequently; this procedure is called *time-stepping.* Having computed the rate of change or residual in the volume, we can then extrapolate that rate for a small time Δt to obtain the value for the state at the time $t + \Delta t$.

- Time-stepping can be compared to observing an unsteady flow at frequent intervals. Most of the times, however, CFD is used to find the steady state solution. The simulation is started with some initial guess for the solution, e.g., uniform flow. Then the time-stepping process is performed until either the residual, that is the rate of change of the states in the cells, has become acceptably small or until quantities of interest such as lift or drag are not changing any longer to a fixed number of digits.

3.6.1 Discretisation on arbitrary grids

Section 3.4 introduced the concept of a regular, structured grid and demonstrated how to discretise the advection equation on such a grid. Chapter 7 will discuss in more detail the pros and cons of the major types of grids, but it is clear that it will be very difficult to fit such regular lattices onto the complex domains of industrial CFD applications.

Modern CFD solvers typically work with *unstructured* grids that do not assume any particular regularity or ordering of the cells in the grid. Typically they also allow a mix of elements, e.g., triangles and quadrilaterals in 2-D, or even control volumes with an arbitrary number of faces.

Fig. 3.12 shows such a grid around an aerofoil that is composed of two different types of elements. Quadrilaterals are used in the boundary layer

[12]There is a large body of literature covering some 30 years of research on how best to formulate these fluxes. This subject shall be reserved for more advanced studies in CFD. Hirsch [2] gives an overview.

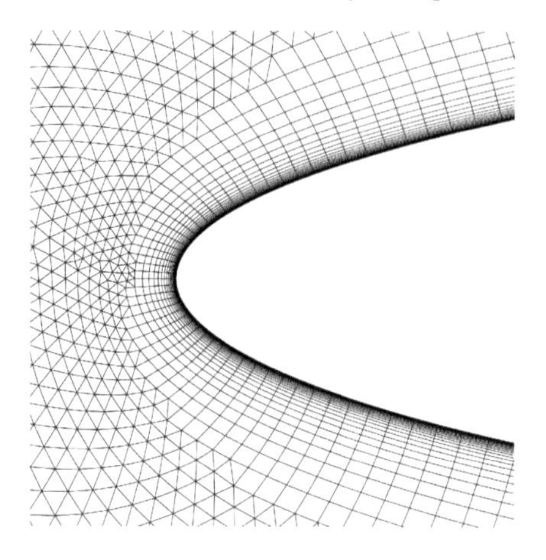

FIGURE 3.12
Unstructured grid around an aerofoil: quadrilateral cells to accurately capture boundary layer effects, triangular elements in the far-field for flexibility.

region which improves the accuracy of capturing the strong gradients normal to the wall. Triangles are used in the far-field, as they allow to rapidly coarsen the mesh where only a lower mesh density is required.

Using arbitrary element types and grid structures does not pose a particular problem to the finite volume method; in fact, Eq. 3.20 applies directly, as long as we appropriately adjust the number of faces k for each control volume. In the discussion of the implementation of the advection equation on the structured grid, we have looped over the faces to compute cell residuals, making use of the fact that fluxes are the same when viewed from either side of the interface. This can be extended to arbitrary unstructured grids by introducing a data structure that tells the solver what cells are either side of each face in the grid. Ch. 7 discusses structured and unstructured grids in more detail.

3.6.2 Transport through an arbitrary face

In the transport equations of fluid flow, the flux balance or residual is equivalent to the rate of change as shown in (3.19). On an arbitrary grid this flux balance is computed as a sum over all the faces of the cell. Fig. 3.13 shows a quadrilateral cell with four faces.

In two- or three-dimensional flow the flux $\mathbf{F}$ between the two faces is a vector, e.g., the mass flux $\mathbf{F}_\rho = \mathbf{v}\rho$ where ρ is the density and $\mathbf{v}$ is the transporting velocity field. Only the component of $\mathbf{F}$ perpendicular to the face k, in other words the component parallel to the face normal $\mathbf{n}_k$, actually transports in or out of the cell, and the transport is proportional to the area

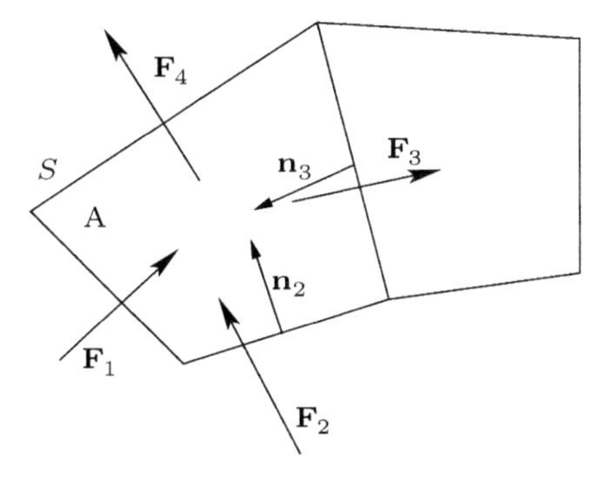

FIGURE 3.13
Interface between two cells with conservative flux between the cells.

s_k of the face k. Hence the net contribution is

$$f_{\perp,k} s_k = \mathbf{F} \cdot \mathbf{n}_k s_k.$$

In the example of an upwinded mass flux $\mathbf{F}_{\perp,k} s_k = \rho \mathbf{v}_3 \cdot n_k s_k$.

Specialising the pure transport equation (3.19) for a control volume wih K straight sides we find

$$\begin{aligned} \int_A \frac{\partial u}{\partial t} dA = \frac{\partial u}{\partial t} \int_A dA &= \oint_S \mathbf{F} \cdot \mathbf{n} \, dS \\ A \frac{\partial u}{\partial t} &= \sum_{k=1}^{K} \mathbf{F}_k \cdot \mathbf{n}_k s_k \\ &= \sum_{k=1}^{K} u \mathbf{v}_k \cdot \mathbf{n}_k s_k \end{aligned} \tag{3.21}$$

where u is the average value of the conserved quantity, e.g., setting $u = \rho$ leads to mass conservation. This, of course, is equivalent to (3.20), except that we allow an arbitrary number K of faces with arbitrary orientation $\mathbf{n}_k$.

3.6.3 The concept of pseudotime-stepping

The explicit form of time-stepping has been introduced for the simple examples in Ch. 3. The more commonly used implicit time stepping is presented in Chapter 4.5. We call *Pseudotime-stepping* the typically used approach in CFD where the time derivative is discretised, but only used as a means to reach the steady state. "Pseudo" refers to the fact that we don't need an accurate value for the time derivative, so we are free to dilate or compress time in order to reach the steady state more quickly. In this section we have a brief look at how to control this pseudo-transient, how we can ensure that the pseudo-unsteady solution is steady enough to stop the iterations.

The flux balance on the right-hand side of (3.9) is called the *residual*. It is this [im]balance that drives the flow since it is equivalent to the rate of change on the left-hand side. The residual is a measure of the unsteadiness of a flow. It is not a measure of error in the solution.

Hence even if only the steady state solution is sought, we still view our method as mimicking the transient from an initial solution to the final steady state one. As an analogy one might think of a wind tunnel experiment. After mounting the model one would turn on the wind tunnel, wait for the flow to become steady and then take any measurements. In the same way, for a steady state CFD solution we need to perform a sufficient number of time-steps for the solution to have become steady, i.e., for the solution to have *converged*.

One way of assessing the steadiness is by monitoring the residuals: if they become small the rate of change becomes small.

At first one could think of running the iteration process until the residuals have been reduced to zero. A zero residual can, however, not be achieved on a digital computer as there is only a finite precision of the numbers in storage. Fluent e.g., uses single precision numbers, they are represented with around 7 significant digits. Other packages use double precision that offers 14 significant digits. Any number resulting from a floating point operation such as multiplication or division will have a *round-off* error, basically a random fluctuation of the 8th digit and beyond due to the finite-precision arithmetic in computers (cf. Sec. 4.2.1). This will mean that a residual can converge to at most *machine zero* of 10^{-7} for single precision or 10^{-14} for double precision computations.

Often this is not achievable due to the numerical stability of the iterative process and the residual only converges 3 or 4 orders of magnitude compared to the residual of the initial solution. Quantities of interest to an engineer, such as e.g., lift and drag of an aerofoil or mass flow through an engine inlet, converge differently from the residual. It is best to monitor the convergence of these quantities directly and stop the computation once the figures are constant to the 3 or 4 digits of accuracy that a CFD solution can provide. It is more practical, however, to provide a criterion to stop the simulation once size of the residual has dropped below a certain threshold. Other quantities can oscillate around the final steady state value and it is difficult to design a stopping criterion based on their steadiness. With experience a user will know what levels of residual correspond to a sufficiently steady quantity of interest.

In cases of periodic flows, e.g., vortex shedding behind a cylinder, the residuals will oscillate and never converge to low values. Convergence problems are also often encountered with detailed CFD solutions, e.g., aerofoil simulations that resolve the bluntness of the trailing edge which sheds vortices. To capture these vortices and their effect on the average flow field is essential for very accurate simulations of flows with high angles of attack.

3.6.4 Time-stepping for compressible flows

The physics of the compressible flow equations are much more complex than those for incompressible flow, but the structure is much simpler. Shown for illustration (not derived) here are the 2-D equations; see a basic textbook for their derivation:

$$\begin{aligned}
\frac{\partial \rho}{\partial t} =& \frac{\partial \rho u}{\partial x} + \frac{\partial \rho v}{\partial y} \\
\frac{\partial \rho u}{\partial t} =& \frac{\partial \rho u^2 + p - \tau_{xx}}{\partial x} + \frac{\partial \rho uv - \tau_{yx}}{\partial y} \\
\frac{\partial \rho v}{\partial t} =& \frac{\partial \rho uv - \tau_{xy}}{\partial x} + \frac{\partial \rho v^2 + p - \tau_{yy}}{\partial y} \\
\frac{\partial \rho E}{\partial t} =& \frac{\partial ((\rho E + p)u - u\tau_{xx} - v\tau_{xy} + q_x)}{\partial x} \\
& \frac{\partial ((\rho E + p)v - u\tau_{xy} - v\tau_{yy} + q_y)}{\partial y}.
\end{aligned}$$

They are closed by an equation of state for the pressure, e.g., the ideal gas equation

$$p = \rho RT.$$

The calculation of the absolute pressure requires to calculate an energy equation so we can compute the temperature T. The compressible Naiver-Stokes equations in two dimensions then have 4 equations for the 4 unknowns of density ρ, x, y-velocities u, v and the energy E.

All equations (bar the simple state equation) have the form of a *transport equation*. A rate of change on the left-hand side comes about due to matter/momentum/energy being transported in or out by the flow and/or being created or destroyed (e.g., viscous friction destroys momentum and generates heat). Given an initial solution, we can advance the solution in time for all components.

The compressible flow equations are strongly coupled through the flow physics; it is not effective to solve each equation separately. The compressible equations are typically solved in a coupled fashion: similar to the example of the advection equation (3.21), each component of the flow vector is iterated in the same way. A simple explicit scheme for the state vector $\mathbf{U} = [\rho, \rho u, \rho v, \rho E]$ of conserved variables in two dimensions and the corresponding flux vector $\mathbf{F}$ could look like

$$A\frac{\partial \mathbf{U}}{\partial t} = \sum_{k=1}^{K} \mathbf{F}_k \cdot \mathbf{n}_k s_k.$$

The equations include the representation of sound waves, small pressure and density variations. This is essential to simulate high-speed flows at Mach numbers above 0.3, or flows where the propagation of sound waves (acoustics) is relevant. Acoustic waves travel at the speed of sound, so the discretisations need to take this speed into consideration for stability and the scaling of artificial viscosity.

In most applications of CFD other than aeronautics, flow speeds are typically very low and the density variations compared to atmospheric are very small. For slower speeds the ratio of the fast sound waves compared to the flow speed of interest becomes very large and compressible discretisations become very ineffective. Using a compressible discretisation for very slow flows would result in very poor convergence to the steady state and excessive amounts of artificial viscosity. For these cases, we need to use discretisations based on the incompressible flow equations instead.

3.6.5 Iterative methods for incompressible flows

As derived in Secs. 2.2.2 and 2.3.4, when neglecting gravity, and in two dimensions, the incompressible Navier-Stokes read

$$0 = \frac{\partial u}{\partial x} + \frac{\partial v}{\partial y} \qquad (2.6)$$

$$\frac{\partial u}{\partial t} = -u\frac{\partial u}{\partial x} - v\frac{\partial u}{\partial y} - \frac{1}{\rho}\frac{\partial p}{\partial x} + \nu\left(\frac{\partial^2 u}{\partial x^2} + \frac{\partial^2 u}{\partial y^2}\right) \qquad (2.30)$$

$$\frac{\partial v}{\partial t} = -u\frac{\partial v}{\partial x} - v\frac{\partial v}{\partial y} - \frac{1}{\rho}\frac{\partial p}{\partial y} + \nu\left(\frac{\partial^2 v}{\partial x^2} + \frac{\partial^2 v}{\partial y^2}\right). \qquad (2.31)$$

The structure of the compressible flow equations allows to use a rather straightforward generalisation from a scalar advection to the set of flow equations. While the physics of incompressible flow is simpler than compressible flow, two aspects make the discretisation of the equations much more difficult:

- The incompressible momentum equations have a similar structure to the compressible ones, but there is no time derivative with the continuity equation. This is no longer a transport equation, but a *constraint*: the *divergence* of the flowfield has to be zero at all times!
- There is no equation to compute the pressure.

One approach to circumvent this is to introduce an artificial compressibility which is made proportional to the rate of change of density $\varepsilon\frac{\partial\rho}{\partial t}$ and hence vanishes in the steady state, recovering the incompressible flow solution. An artificial equation of state then needs to be formulated for the pressure. It is in practice difficult to select an appropriate value of ε, with a poor choice adversely affecting the convergence rate, and hence this approach is not widely used.

The most popular method uses an alternative approach by combining continuity and momentum equations to obtain an equation for the pressure, the *pressure correction equation*, hence called the pressure-correction or fractional step methods. The principle of this approach exploits the fact that pressure and velocity are intricately coupled, e.g., a change in pressure field will deviate the fluid particles, hence change the velocity field. The Bernoulli equation captures this concisely for inviscid, incompressible flow: the total pressure, which is the sum of static pressure and dynamics pressure, has to be constant. The idea is hence to solve transport equations for the velocities using some guess for the pressure field. This will typically result in an updated velocity field that does not satisfy continuity. The pressure correction then quantifies the modification of the pressure field that is required in order to make the flowfield divergence-free.

To derive the pressure correction equation we will use the more compact vector notation of the equations. Recall the gradient operator ∇ applied to a scalar u:

$$\nabla u = \left[\frac{\partial u}{\partial x}, \frac{\partial u}{\partial y}, \frac{\partial u}{\partial z}\right]^T .$$

Applied to a vector field $\mathbf{u} = [u, v, w]^T$ we have

$$\nabla \mathbf{u} = \begin{bmatrix} \frac{\partial u}{\partial x} & \frac{\partial u}{\partial y} & \frac{\partial u}{\partial z} \\ \frac{\partial v}{\partial x} & \frac{\partial v}{\partial y} & \frac{\partial v}{\partial z} \\ \frac{\partial w}{\partial x} & \frac{\partial w}{\partial y} & \frac{\partial w}{\partial z} \end{bmatrix} .$$

We can then write the 3 momentum equations in x, y, z compactly as

$$\frac{\partial \mathbf{u}}{\partial t} = -\mathbf{u}\nabla\mathbf{u} - \frac{1}{\rho}\nabla p + \nu\nabla^2\mathbf{u}. \quad (3.22)$$

Recall also the definition of the divergence operator $\operatorname{div}(\mathbf{u})$ on a vector field $\mathbf{u}$ with components u, v, w:

$$\operatorname{div}(\mathbf{u}) = \nabla\cdot\mathbf{u} = \frac{\partial u}{\partial x} + \frac{\partial v}{\partial y} + \frac{\partial w}{\partial z}$$

The divergence is the incompressible mass balance or continuity.

The momentum equation is an equation for the rate of change of the velocity field $\frac{\partial \mathbf{u}}{\partial t}$. If we ensure that the initial solution $\mathbf{u}^0$ is divergence-free, and if we ensure that each update $\Delta t\frac{\partial \mathbf{u}}{\partial t}$ is divergence-free, then also the solution $\mathbf{u}^n$ at timelevel n will be divergence-free. Taking the divergence of the momentum equation (3.22) with some intermediate velocity $\mathbf{u}^*$ and some intermediate pressure p^*, we find the divergence error Err_{Div} to be

$$Err_{Div} = \nabla\cdot\left(\frac{\partial \mathbf{u}^*}{\partial t}\right) = \nabla\cdot\left(-\mathbf{u}^*\nabla\mathbf{u}^* - \frac{1}{\rho}\nabla(p^*) + \nu\nabla^2\mathbf{u}^*\right).$$

To make this error zero, we add a correction to the pressure field p^* to obtain the correct divergence-free pressure field p as $p = p^* + \Delta p$:

$$0 = \nabla\cdot\left(\frac{\partial \mathbf{u}^*}{\partial t}\right) = \nabla\cdot\left(-\mathbf{u}^*\nabla\mathbf{u}^* - \frac{1}{\rho}\nabla(p^* + \Delta p) + \nu\nabla^2\mathbf{u}^*\right).$$

The pressure correction equation for the pressure field that would result in a velocity change $\frac{\partial \mathbf{u}}{\partial t}$ with zero divergence to satisfy continuity:

$$\nabla \cdot (\nabla(p^* + \Delta p)) = \nabla^2(p^* + \Delta p) = \rho \nabla \cdot \left(-\mathbf{u}^* \nabla \mathbf{u}^* + \nu \nabla^2 \mathbf{u}^*\right)$$
$$\nabla^2 p = \rho \nabla \cdot \left(-\mathbf{u}^* \nabla \mathbf{u}^* + \nu \nabla^2 \mathbf{u}^*\right). \qquad (3.23)$$

The form of this equation is a Poisson equ. as introduced in Sec. 2.4.3.

The pressure correction to be applied to the momentum equations to obtain a divergence-free velocity field is then

$$\nabla p = \nabla(p^* + \Delta p). \qquad (3.24)$$

Applying this correction in turn then does not satisfy the momentum equations, so a number of pressure correction iterations have to be performed.

The pressure correction equation is not just some neat mathematical trick, but very physical: if we pump fluid into a C.V. in compressible flow, the pressure will rise, as the compressibility acts like a spring. In incompressible flow we cannot increase mass in the cell; the divergence has to remain zero: the pressure would rise infinitely as fluid is incompressible, and the spring is infinitely stiff. The pressure correction equation quantifies how much the pressure has to change in each control volume in order to repel or attract flow into or out of the volume in order to satisfy continuity.

3.6.6 The SIMPLE scheme

The SIMPLE scheme[13] is the standard scheme for incompressible flow. For each iteration or 'time-step', the following is calculated:

1. Guess a pressure field $p*$ (or use p^n from the last iteration).
2. Solve the momentum equations (3.22) for the velocity field u^*, based on the guessed pressure $p*$, which will not be divergence-free.
3. Solve the pressure correction equation (3.23) for Δp.
4. If Δp is "small enough" then the equations are solved, exit, else
5. Correct the pressure p^* using (3.24).
6. Correct the velocities u^* with the modified pressure gradient in (3.22).
7. Go to step 2.

The resulting systems are solved implicitly, i.e., a large system of equations arises that couples the unknowns at all mesh points. Implicit methods are discussed in more detail in Sec. 4.5. The SIMPLE scheme avoids solving implicit systems of equations for the entire system of u, v, w, p. Instead, it only solves four much smaller systems for u, then v, then w, then p. The fully

[13]Semi-Implicit Method for Pressure-Linked Equations.

coupled compressible scheme (or alternative, fully coupled incompressible formulations not presented here) need to solve one large system for all equations simultaneously.

SIMPLE typically converges very fast and efficiently, but for industrial flows in complex geometries and with marginal flow stability, convergence may stall. Typically the residual of the pressure correction equation, which is a difficult equation to solve, only converges 2-4 orders of magnitude. There are other pressure-correction formulations with better accuracy and stability, but higher cost, e.g., SIMPLEC, SIMPLER, PISO, PIMPLE. Discussions of those can be found e.g., in Ferziger [3].

3.7 Exercises

3.1 Implement the first-order upwind scheme for the one-dimensional linear advection equation as discussed in Sec. 3.1.2. Use either a spreadsheet or Matlab. Use a mesh of $N = 64$ mesh points, $a = 1$, inflow boundary condition of $u = 0$, and an initial solution of $u = 0$ for all nodes, except for nodes $1 \leq i \leq 4$ where $u = 1$ and define hence the width of the block profile as equal to 5 nodes. Run enough time-steps to transport the solution by 10 times the width of the block profile, i.e., the solution of node $i = 4$ should have been transported by $a = 1$ to node $i = 54$. Run your "code" with $\nu = .5$ and verify your implementation against what you expect to see.

What is the effect of choosing different CFL numbers between $0 \leq \nu \leq 1$ on the accuracy of the solution? What happens if you use $\nu < 0$ or $\nu > 1$?

3.2 Show that the finite difference discretisation of (3.15) is conservative.

3.3 Derive the five-point discretisation for the second derivative described in Sec. 3.3.1.

3.4 Use this five-point molecule and explicit time-stepping to implement a discretisation for the heat equation in Matlab (or similar). Use a square domain $0 \leq x, y \leq 1$ with $N = 5$ nodes in each direction. As boundary condition, apply a fixed temperature value at the borders as $T(x, y) = 0$ at the left $(x = 0)$, top $(y = 1)$ and right $(x = 1)$ boundaries; apply $T(x) = |x - 0.5|$ at the bottom, $y = 0$.

Verify your solution against expected behaviour (if your solution becomes oscillatory, reduce the time-step).

3.5 Using the code from Ex. 3.4 with the explicit time-stepping, adjust the time-step until you determine the stability limit. Then refine the grid to $N = 10$ and $N = 20$. How are time-step and mesh size related to each other?

4

Analysis of discretisations: accuracy, artificial viscosity and stability

The finite volume method is difficult to analyse in the form given in (2.12), but we have seen in Sec. 3.1.6 that any finite volume method can be turned into a finite difference method which allows to analyse the properties of the method more easily.

We have already encountered some arbitrary choices for approximating derivatives in Sec. 3.1.1. Let us have a closer look at some of the possible choices and analyse their properties.

4.1 Forward, backward and central differences

Consider having to approximate a first derivative such as $\frac{\partial u}{\partial x}$ for the linear advection equation (2.4.1) on a sequence of discrete points as shown in Fig. 4.1. A straightforward discretisation would be to use the mathematical definition of the derivative, and take the limit as far as possible, i.e., the mesh width h:

$$\frac{\partial u}{\partial x} = \lim_{\delta x \to 0} \frac{u(x+\delta x) - u(x)}{\delta x} \approx \frac{u_{i+1} - u_i}{h}. \tag{4.1}$$

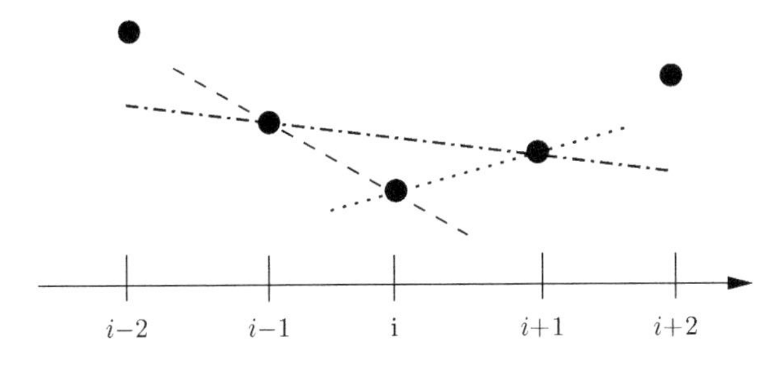

FIGURE 4.1
Approximation of the first derivative at point i on a 1-D mesh of discrete points: forward difference (dotted), backward difference (dashed), central difference (dash-dots).

This is the dotted line in Fig. 4.1. Since the pattern of the points involved, also called the *stencil* or the *molecule*,[1] is biased towards larger i, it is called a *forward difference*. Without particular knowledge of the behaviour of the underlying partial differential equation that is being solved, there is no particular reason why this should be a better choice than any other. We have already seen in Sec. 3.1 that it may be a good idea to bias the molecule into the "wind" of transport; hence, if that direction is directed from node $i+1$ to node i, the molecule could also be biased in the other direction, the dashed line in Fig. 4.1:

$$\frac{\partial u}{\partial x} = \lim_{\delta x \to 0} \frac{u(x) - u(x-\delta x)}{\delta x} \approx \frac{u_i - u_{i-1}}{h}. \tag{4.2}$$

This is a *backward difference*, the dashed line. Another possibility is to make the molecule symmetric, resulting in the *central difference* over twice the mesh width, shown in the line with dash-dots:

$$\frac{\partial u}{\partial x} = \lim_{\delta x \to 0} \frac{u(x+\delta x) - u(x-\delta x)}{2h} \approx \frac{u_{i+1} - u_{i-1}}{2h}. \tag{4.3}$$

If the limit $\lim h \to 0$ can be taken, all 3 variants are equivalent, but in the discretisation the limit can only be pushed as small as $\delta x = h$ and the 3 discretisations lead to very different approximations to the derivative. As can be observed in Fig. 4.1 for a function with a positive second derivative, the forward difference overestimates the slope, the backward difference underestimates it. The central difference gives the best result. In the next section we will analyse why this is so.

4.2 Taylor analysis: consistency, first- and second-order accuracy

The mathematical analysis for this example is actually very simple. If it is assumed that the continuous function which is represented discretely is continuous and continuously differentiable, then it can be expanded in a convergent Taylor series around each mesh point. The neighbouring point values $u(x+\delta x)$ can be replaced by a Taylor expansion around $u_i = u(x)$. For convenience of writing, let again the mesh width be $h = \delta x$

$$u(x + \delta x) = u(x + h) = u(x) + h\frac{\partial u}{\partial x} + \frac{h^2}{2}\frac{\partial^2 u}{\partial x^2} + O(h^3). \tag{4.4}$$

$O(h^3)$ refers to terms proportional to h^3 which have been truncated from the series.

[1] Imagine the mesh points involved being connected like ball-and-stick models of the chemistry kit.

Using this expansion the forward approximation Eq. 4.1 of the first derivative can be rewritten:

$$\left.\frac{\partial u}{\partial x}\right|_{discr} = \frac{u_{i+1} - u_i}{h} \tag{4.5}$$

$$= \frac{(u_i + h\frac{\partial u}{\partial x} + \frac{h^2}{2}\frac{\partial^2 u}{\partial x^2} + O(h^3)) - u_i}{h} \tag{4.6}$$

$$= \frac{\partial u}{\partial x} + \frac{h}{2}\frac{\partial^2 u}{\partial x^2} + O(h^2). \tag{4.7}$$

The analysis reveals four aspects:

- The leading term is $\frac{\partial u}{\partial x}$. There is no factor to scale it; the estimation yields the right value. Also, all error terms in addition to $\frac{\partial u}{\partial x}$ are proportional to the mesh width h. That is, as the mesh is refined, the errors are being reduced. In the — practically impossible — limit of infinitely small mesh width the errors vanish. This property is called *consistency* and is essential to a numerical discretisation. With an inconsistent scheme one cannot expect to find the correct solution as the mesh is refined.
- The discrete approximation does reproduce a first derivative correctly, albeit with a *truncation error* of $\frac{h}{2}\frac{\partial^2 u}{\partial x^2} + \frac{h^3}{6}\frac{\partial^3 u}{\partial x^3} + \cdots$. Since h is a very small quantity, $h \ll 1$, its higher powers are negligible compared to h. The notation $O(h^2)$ is used to collect all terms of the Taylor expansion that are proportional to h^2 or higher powers and are neglected. The truncation error can then compactly be written as $\frac{h}{2}\frac{\partial^2 u}{\partial x^2} + O(h^2)$.
- The leading term of the truncation error is proportional to the first power of the mesh width h. Since h is small, h^2 is much smaller and the leading term is the largest error term. The forward difference is said to be *first-order accurate* as its truncation error scales with h^1.
- Considering that $\frac{\partial^2 u}{\partial x^2}$ is positive in our example, it is now easily understood why the forward difference over-predicts the derivative in our example with positive second derivative; the leading error term has a positive sign.

Similarly, one can expand around u_i with $-h$ for the backward difference.

$$u'_{discr} = \frac{u_i - u_{i-1}}{h} \tag{4.8}$$

$$= \frac{u_i - (u_i - h\frac{\partial u}{\partial x} + \frac{h^2}{2}\frac{\partial^2 u}{\partial x^2} + O(h^3))}{h} \tag{4.9}$$

$$= \frac{\partial u}{\partial x} - \frac{h}{2}\frac{\partial^2 u}{\partial x^2} + O(h^2). \tag{4.10}$$

It can be concluded that the backward difference is also first-order accurate, but the leading error term has the opposite sign, explaining the under-prediction of the derivative in the example.

For the central difference approximation one finds

$$u'_{discr} = \frac{u_{i+1} - u_{i-1}}{2h} \tag{4.11}$$

$$= \frac{1}{2h}\left[\left(u_i + h\frac{\partial u}{\partial x} + \frac{h^2}{2}\frac{\partial^2 u}{\partial x^2} + \frac{h^3}{6}\frac{\partial^3 u}{\partial x^3} + O(h^4)\right) - \right. \tag{4.12}$$

$$\left.\left(u_i - h\frac{\partial u}{\partial x} + \frac{h^2}{2}\frac{\partial^2 u}{\partial x^2} - \frac{h^3}{6}\frac{\partial^3 u}{\partial x^3} + O(h^4)\right)\right] \tag{4.13}$$

$$= \frac{\partial u}{\partial x} + \frac{h^2}{3}\frac{\partial^3 u}{\partial x^3} + O(h^3). \tag{4.14}$$

That is, the first-order error terms cancel and the expression becomes *second order accurate*; the truncation error is proportional to h^2.

4.2.1 Round-off errors

Taylor analysis uses exact arithmetic to analyse the error of a discretisation. The approach assumes that we can make h arbitrarily small and still correctly evaluate the derivative. Computers work with finite precision arithmetic. Floating point numbers are represented using a finite number of bytes for each number, 32 bytes for a *single precision*, 64 bits for a *double precision* number. The computer represents floating point numbers in exponential format. With some bits used for the sign, the exponent and its sign, and some other flags, a single precision number typically has 7 significant digits of accuracy — very similar to a typical pocket calculator —, while a double precision number has 14 significant digits.

Let us consider as an example to compute using single-precision arithmetic a pressure gradient on a mesh of width h. Let us assume that the gradient is $\frac{\partial p}{\partial x} = 5555\,\text{Pa/m}$, and that cell i has a near-atmospheric pressure of $100\,\text{kPa}$; cell $i+1$ then has a value of $p_i + h\frac{\partial p}{\partial x} = h \cdot 5555\,\text{Pa/m}$. Table 4.1 presents the finite difference approximation to the gradient and the error for various values of h. The problem stems from the fact that both p_i and p_{i+1} are large values compared to the difference Δp between them. For the same gradient the difference in pressures rapidly becomes to small to be accurately represented with the 7 digits of accuracy available in this single precision example. The division by a small h then amplifies the error.

In incompressible flow the pressure is an additive constant; we can choose, e.g., to work with a reference pressure of 1 bar or a gauge pressure of 0 Pa. The latter choice removes the problem of round-off errors as the pressure difference now is of a similar order of magnitude as the actual pressure. Provided that additive constants in initial and boundary, as well as any normalising constants are appropriately chosen, round-off errors are not relevant at the mesh widths that are achievable in practical cases.

h	p_i	p_{i+1}	Δp	$\frac{\partial p}{\partial x}\|_{fwd}$	err[%]
1e+00	100000.0	105555.0	5555.0	5555.0	0.0
1e-01	100000.0	100555.5	555.5	5555.0	0.0
1e-02	100000.0	100055.6	55.6	5560.0	0.1
1e-03	100000.0	100005.6	5.6	5600.0	0.8
1e-04	100000.0	100000.6	0.6	6000.0	8.0
1e-05	100000.0	100000.1	0.1	10000.0	80.0
1e-06	100000.0	100000.0	0.0	0.0	100.0
1e-07	100000.0	100000.0	0.0	0.0	100.0

TABLE 4.1
Round-off error for single precision arithmetic.

4.2.2 Order of accuracy and mesh refinement

For a numerical calculation this has important implications. Consider having an initial mesh on which the results are found to be unsatisfactory and one decides to use a finer mesh. Halving the mesh width h and using a first-order accurate scheme, one will find that the truncation errors are approximately halved as well. However, if a second order accurate scheme is used, halving the mesh size will reduce the truncation error approximately to a fourth!

Note that the Taylor expansion is only valid for smooth functions; hence Taylor analysis allows the analysis of the truncation error of smooth functions. It does not address errors by modelling or by inadequate temporal convergence[2] of the residuals due to too few time-steps. Also this analysis does not apply where the solution is discontinuous, as e.g., in shocks or where the solution is not smooth enough if the mesh is really coarse.

Finite volume methods do not employ the simple discretisations shown above, but as we have seen we can always reformulate a finite volume method as a finite difference method and then analyse the properties using Taylor expansion. In the advanced CFD packages the user has a choice between first- and second-order accurate discretisations. Hence choosing a first-order method the truncation error will be proportional to the mesh width h; in the second-order method it will be proportional to the square of the mesh width h^2.

It needs to be pointed out that the second-order discretisations available in CFD codes are not based on the central difference; this would result in an unstable discretisation as shown in Sec. 4.3.1. Instead they are based on a discretisation over 5 points with an upwind bias in the weights of the *molecule*.

[2]We need to distinguish between the temporal convergence of the residuals to zero (or near enough) by running enough time-steps and the mesh convergence of the solution toward the exact solution by repeatedly refining the mesh.

Secs. 4.3.3 and 4.3.4 discuss this in more detail. The first-order discretisations typically used are indeed similar to the first-order upwind scheme of Sec. 3.1.4 using constant states over the cells and backward differences to approximate the fluxes.

Very few packages based on finite elements (not dicussed here) are available for specialised applications, such as e.g., polymer flows. For a second-order finite element method the same property holds: halving the mesh size reduces the truncation error to a quarter. Finite element methods are less popular in CFD, mainly due to the fact that finite volume methods offer a well-developed methodology to retain monotonicity. Finite element codes for fluids are available for specialist applications and as research codes. The main advantage of finite element discretisations is that they can offer a much higher order of accuracy than the second-order typical for finite volume methods.

4.3 Stability, artificial viscosity and second-order accuracy

We have already encountered the linear advection equation in Sec. 2.4.1 and have seen possible discretisations of the equation, the first-order upwind scheme in its finite difference (3.3) and its finite volume form (3.10):

$$u_i^{n+1} = u_{i-1}^n - \frac{a\Delta t}{h}(u_i^n - u_{i-1}^n) \tag{3.3}$$

$$u_i^{n+1} = u_{i-1}^n - \frac{\Delta t}{h}(f_i^n - f_{i-1}^n). \tag{3.10}$$

To analyse the quality of the solution in detail, we could use a frame of reference that moves with the advection speed a. In this frame our initial profile, e.g., a step, should be exactly preserved and stationary.

The first-order upwind solution of the step profile has been shown in Fig. 3.4. The solution does not exhibit overshoots. The evolution is *monotonic*, i.e., the solution at the new time-step is bounded by the extrema of the old time-step.[3] We can also observe that the initially sharp discontinuity gets smeared, even though there is no diffusive or viscous term in the advection equation. The first-order upwind method evidently has a large diffusive error.

Can we devise a scheme with better accuracy and less smearing? We have also encountered the central difference equation (4.3) which is second-order accurate. If we replace in the space derivative the first-order backward difference with a second-order central difference, we obtain

$$u_i^{n+1} = u_{i-1}^n - \frac{a\Delta t}{2h}(u_{i+1}^n - u_{i-1}^n). \tag{4.15}$$

[3]This is further discussed in Sec. 4.5.1.

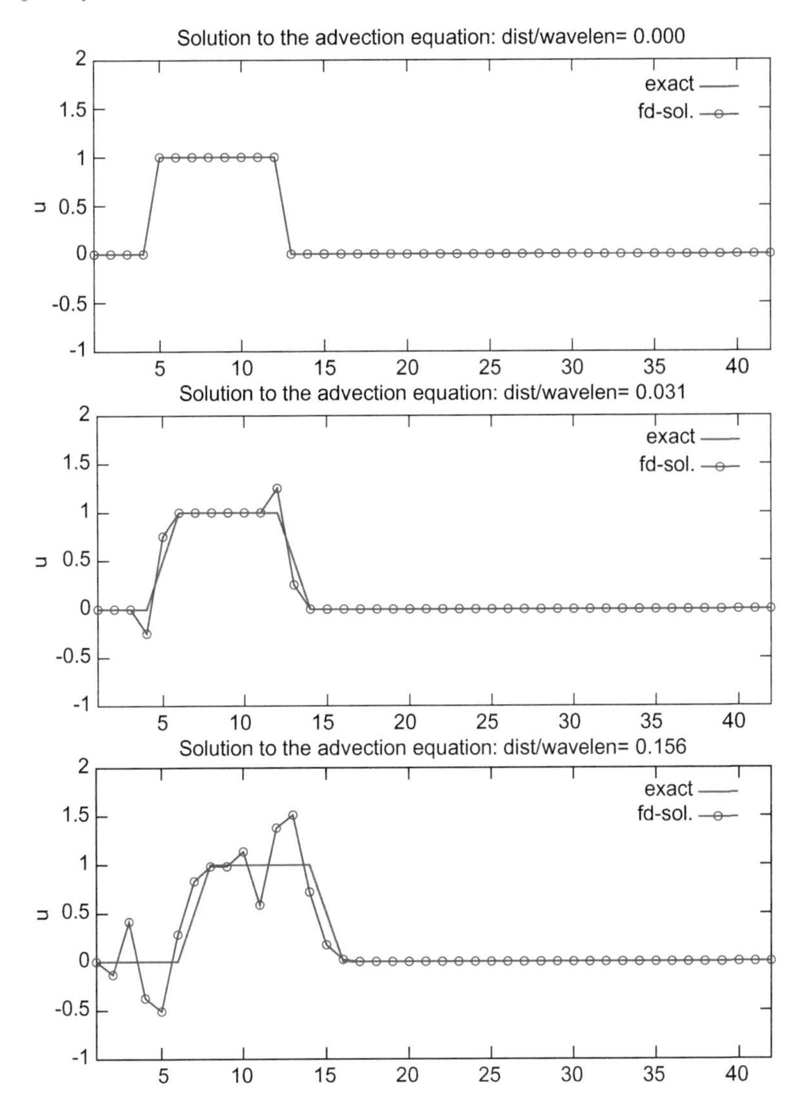

FIGURE 4.2
Advection of a block profile, central space discretisation. Initial solution (top), after 1 iteration with CFL=0.5 (middle), after 2 iterations (bottom).

This scheme is called the *FTCS scheme*, since the discretisation is forward in time, central in space. Figure 4.2 shows what happens with this scheme, again using a Courant number of $\nu = 0.5$. Note the changed scale for u compared to Fig. 3.4.

One can observe that the first iteration creates *overshoots*. The following iterations make the situation even worse; the error is amplified dramatically.

The amplitude of the oscillations will grow until the values become too big to be represented numerically. The simulation will *blow up*.

The central discretisation shown here is unstable, i.e., there are error modes in the solution whose amplitude is increased at every iteration, leading to exponential growth. Clearly, a prerequisite for a CFD discretisation is that it is *stable*, i.e., no modes, whether error modes or not, are amplified.

4.3.1 Artificial viscosity

Can we stabilise the central scheme? We could add a small amount of non-physical or *artificial viscosity* (A.V.) to suppress the amplification of error modes observed with the central scheme of Eq. 4.15. This is very similar to the viscous term in the momentum equation (2.33), except that the second derivative is multiplied with a small factor ε rather than the molecular viscosity μ:

$$\frac{\partial u}{\partial t} + a\frac{\partial u}{\partial x} = \varepsilon\frac{\partial^2 u}{\partial x^2}. \tag{4.16}$$

Discretisation of a second derivative has been discussed for the heat equation in Sec. 3.3.1. This discretisation of the second derivative is second-order accurate; the proof is in Exercise 4.1. The discretisation of (4.16) using the forward difference in time, the central difference in space and the second difference (3.16) for the viscous term is

$$\begin{aligned}
\frac{u_i^{n+1} - u_i^n}{\Delta t} + a\frac{u_{i+1}^n - u_{i-1}^n}{2h} &= \varepsilon\frac{u_{i+1}^n - 2u_i^n + u_{i-1}^n}{h^2} \\
u_i^{n+1} &= u_i^n - \frac{\nu}{2}(u_{i+1}^n - u_{i-1}^n) + \frac{\varepsilon}{h^2}(u_{i-1}^n - 2u_i^n + u_{i+1}^n) \\
&= (1 - \frac{2\varepsilon}{h^2})u_i^n + (\frac{\nu}{2} + \frac{\varepsilon}{h^2})u_{i-1}^n + (-\frac{\nu}{2} + \frac{\varepsilon}{h^2})u_{i+1}^n.
\end{aligned}$$

How large or small do we have to choose ε? If we let $\varepsilon = h^2\nu/2$, i.e., $\varepsilon/h^2 = \nu/2$, then

$$\begin{aligned}
u_i^{n+1} &= (1-\nu)u_i^n + (\frac{\nu}{2} + \frac{\nu}{2})u_{i-1}^n + (-\frac{\nu}{2} + \frac{\nu}{2})u_{i+1}^n \\
&= (1-\nu)u_i^n + \nu u_{i-1}^n,
\end{aligned}$$

which is exactly the first-order upwind scheme which we have used in Eq. 3.3. Hence there are two equivalent viewpoints to obtain the first-order upwind scheme.

On the one hand we can use a discretisation of the advective term that is non-symmetrically biased to take in information from 'upwind', resulting in a stable scheme which has a stabilising viscosity built into it by the choice of discretisation. On the other hand we can use a symmetric 'central' discretisation and explicitly add a stabilising dissipation. A particular choice of coefficient for the artificial viscosity results in the elimination of the downstream node from the update formula which hence becomes the first order upwind scheme. Other choices, of course, are possible, but it is not the aim here to discuss them all. As already pointed out, the second-order discretisations in modern CFD codes do not use the central difference for the advective term.

This artificial viscosity is welcome in that it stabilises the iterative time-stepping process. Moreover, not only does it prevent any modes from growing, it also dampens down any error modes and helps us to reach the steady state more quickly. But there is also a downside. The numerical scheme can't distinguish between modes that are actually physical and should be preserved as part of the solution and modes that are errors e.g., from the initial solution. The artificial viscosity is applied to all of them indiscriminately. As we have already observed, there will be viscous artefacts (smearing) in a solution that should be inviscid, hence have no smearing. The artificial viscosity results in an error in our solution.

We have added a second-difference term which is second-order accurate $O(h^2)$, but multiplied with ν which is proportional to h, $\nu = O(h)$; hence, we have added a first-order error term. If we find that the artificial viscosity is too large, we can refine the mesh. Halving the mesh width h will result in half the artificial viscosity in this first-order scheme.

When we choose a second-order accurate scheme in our CFD solver, we actually add a second-order artificial viscosity. Section 4.3.3 looks at how we can achieve second-order accuracy and stability, Sec. 4.3.4 looks at the now standard way to ensure monotonicity. With a second-order formulation for the artificial viscosity, the gain in accuracy will be significant as the artificial viscosity in this case scales with h^2 rather than h. But as a consequence, cutting the mesh width h in half will reduce the artificial viscosity by a factor of 4. The solution is less stable and will take more iterations to converge to the steady state as error modes are not damped down as strongly as would be the case with a first-order method.

Ensuring adequately low levels of artificial viscosity is one of the key difficulties encountered when running CFD simulations. Determining the level of artificial viscosity is a matter of experience; judging its effect on the solution is even more difficult. Ideally one would want to refine the mesh to a size that reduces the artificial viscosity such that it is much smaller than other errors incurred. This state is called *mesh convergence*; further refinement does not change the results of the simulation. In 3-D simulations the required mesh sizes are in general not feasible, however, and the user has to develop good judgement.

4.3.2 Artificial viscosity and finite volume methods

Recall that the hallmark of a finite volume method is that the state values in a cell are representative of the average state value in the cell. First-order finite volume methods simply assume that the state is constant over the cell, i.e., the

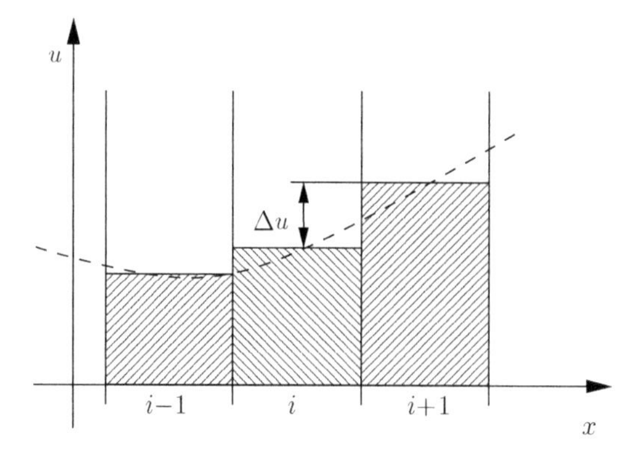

FIGURE 4.3
First-order finite volume method.

solution is *piecewise* constant over the mesh (see Fig. 3.6 for an illustration). A sample variation in 1-D is shown in Fig. 4.3.

Unless the approximated function is constant, the first-order finite volume representation produces jumps at the cell interfaces. Fig. 4.3 shows a jump of Δu between cells i and $i+1$. The jump can be quantified to leading order by using a Taylor expansion:

$$\begin{aligned} \Delta u &= u_{i+1} - u_i \\ &= u_i + h\frac{\partial u}{\partial x} + O(h^2) - u_i \\ &= h\frac{\partial u}{\partial x} + O(h^2). \end{aligned} \tag{4.17}$$

The leading term of the jump is proportional to the first derivative and is of first-order. The errors due to the artificial viscosity must not dominate the errors in the solution. They must not be larger than the truncation error. Hence, the artificial viscosity needs to be at least[4] of first-order. A well-designed artificial viscosity is made to be proportional to the jump Δu, hence also proportional to the first derivative. The first-order upwind discretisation includes an artificial viscosity of precisely that form.

The analysis in Eq. 4.17 also shows that the artificial viscosity can be viewed as a form of truncation error. However, not all truncation errors are dissipative.[5]

Figure 4.4 shows a second-order finite volume approximation of the same function. The second-order accurate finite volume method assumes that the function varies linearly over the cell; the solution is *piecewise linear*. It can be

[4]Least meaning here of that order or higher, hence of smaller magnitude.

[5]There are also dispersive errors that affect the correct speed of waves travelling in the solution. These are relevant to strongly unsteady simulations such as LES or DNS, but are not considered here.

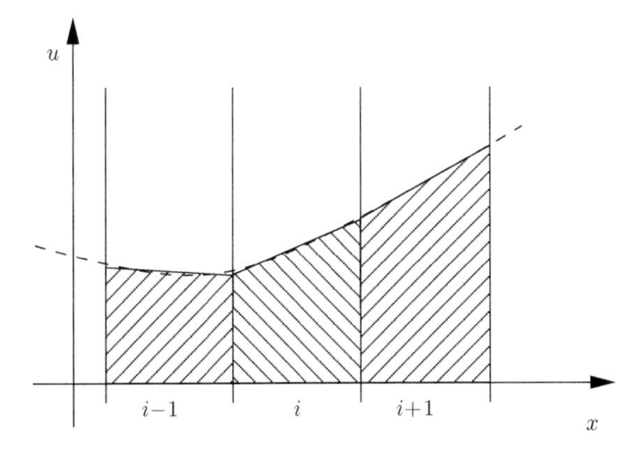

FIGURE 4.4
Second-order finite volume method.

observed that this approximation is much better. The jumps at the interfaces are barely discernible. The state value at the right end of the cell is

$$u_i^+ = u_i + \frac{h}{2}\frac{\partial u}{\partial x}, \tag{4.18}$$

where some approximation of the first derivative is taken to extrapolate by half the mesh width to the right. A Taylor expansion produces

$$\begin{aligned}
\Delta u &= u_{i+1}^- - u_i^+ && (4.19)\\
&= \left(u_{i+1} - \frac{h}{2}\frac{\partial u}{\partial x}\right) - \left(u_i + \frac{h}{2}\frac{\partial u}{\partial x}\right) && (4.20)\\
&= \left(\left(u + h\frac{\partial u}{\partial x} + \frac{h^2}{2}\frac{\partial^2 u}{\partial x^2} + O(h^3)\right) - \frac{h}{2}\frac{\partial u}{\partial x}\right) - \left(u + \frac{h}{2}\frac{\partial u}{\partial x}\right) && (4.21)\\
&= \frac{h^2}{2}\frac{\partial^2 u}{\partial x^2} + O(h^3), && (4.22)
\end{aligned}$$

confirming that the jump in the second-order finite volume method is second-order accurate.

Ideally the artificial viscosity should be proportional to the jump Δu, making it second-order. This would produce an exact solution for linear functions as the jumps then vanish. For typical discretisations designed for good computational efficiency, this is only the case for regular orthogonal grids. However, for practical purposes it is legitimate to consider that the numerical viscosity scales with the jump at the interfaces.

An important aspect to realise is that the artificial viscosity is not only proportional to h^p for a p-th order accurate discretisation, but also proportional to the p-th derivative. If we want to avoid spots of extreme error[6] we

[6] Sec. 7.5 introduces the idea that in the non-linear flow equations an objective function

need to ensure that the product $h^p \frac{\partial^p u}{\partial x^p}$ remains small enough. For practical purposes we can just consider the product of mesh width times first gradient, $h\frac{\partial u}{\partial x}$ or how large the jump is between the piecewise constant states. In two or three dimensions we have to, of course, also consider y- and z-directions in the same way.

Poor mesh quality also can increase the level of artificial viscosity in a solution. Section 7.1 will discuss mesh quality in more detail.

To keep artificial viscosity small the user has hence three choices. He/she should

- use second-order accurate methods rather than first-order ones,
- use a suitable mesh width that is small where the solution varies strongly, hence the jumps between cells are larger, and use a larger mesh width where the flow is nearly uniform,
- maintain good mesh quality.

The two latter, mesh-related aspects are discussed in more detail in Sec. 7.1.

4.3.3 Stable second-order accurate discretisations for CFD

The FTCS scheme (4.15) was an attempt to find a second-order accurate discretisation for the linear advection equation (2.36), but it is not stable as we have seen. As an alternative approach to obtain a second-order accurate, but stable scheme we could start from the first-order upwind scheme (3.10) which is based on the integral form of the advection equation (3.5),

$$\int_{t^n}^{t^{n+1}} \int_{x-\frac{1}{2}}^{x+\frac{1}{2}} \left(\frac{\partial u}{\partial t} + \frac{\partial f}{\partial x} \right) dx dt = 0, \tag{3.5}$$

presented in Sec. 3.1.4.

We found that using an upwind representation of the fluxes, $f_{i-\frac{1}{2}} = f_{i-1}$, for $a > 0$ could lead to a stable scheme, but the scheme is only first-order accurate and produces a large amount of artificial viscosity. This is due to the scheme using a piecewise constant representation for the data within each finite volume, $u(x) = u_i$ for $x_{i-\frac{1}{2}} \le x \le x_{i+\frac{1}{2}}$ where x_i is the coordinate of the centre of the volume i, and $x_{i\pm\frac{1}{2}}$ are the coordinates of the volume or cell interfaces, as shown in Fig. 4.3.

The representation of the function u can be improved by using a piecewise linear representation as shown in Fig. 4.4. Within each volume, the data then vary as

$$u(x) = u_i + (x - x_i) \left. \frac{\partial u}{\partial x} \right|_i \quad \text{for} \quad x_{i-\frac{1}{2}} \le x \le x_{i+\frac{1}{2}}. \tag{4.23}$$

Note that the choice of the gradient $\partial u/\partial x|_i$ in volume i does not affect

such as lift or drag will be very sensitive to some areas of flow, and much less so to others. The formal discussion of this aspect is beyond the remit of this book, but the user will develop this understanding for the cases she/he is familiar with from practice.

conservation: since the linear and constant representations both have the same value at the centre of the volume, the integral over the volume is identical. A symmetric approach to approximate the gradient would be to use a central difference,

$$\left.\frac{\partial u}{\partial x}\right|_i = \frac{u_{i+1} - u_{i-1}}{2h},$$

which makes the slope in the volume the same as the secant through the two neighbouring states. The linearly reconstructed value at the right interface $i+1$ from within volume i is hence

$$u^-_{i+1} = u_i + (x - x_i)\left.\frac{\partial u}{\partial x}\right|_i = u_i + \frac{h}{2}\frac{u_{i+1} - u_{i-1}}{2h} = u_i + \frac{u_{i+1} - u_{i-1}}{4}$$

where the minus sign in the superscript of u^-_{i+1} indicates that this value is found on the left, or negative, side of the interface. As discussed in Sec. 4.3.2, finite volume methods with linear reconstruction still exhibit jumps at the interfaces. We will expect that the state value extrapolated from the centre of volume $i+1$ to the same interface $i+\frac{1}{2}$ will be different.

Upwinding for $a > 0$ means the flux $f_{i+\frac{1}{2}}$ is based on the state u_i to the left of the interface. In the case of piecewise constant data as used in the first-order upwind scheme, the flux also is piecewise constant. In the context of the advection equation constant in space is equivalent to the fluxes of the first-order upwind scheme being constant in time since the solution profile is transported in time with speed a. We exploited this property when simplifying (3.5) to (3.6).

In the case of a piecewise linear data representation we have to take into account the variation of the flux over the time interval Δt. The average flux over the time interval can be computed using the trapezium rule, i.e., basing the flux on the linear representation of u found at the interface $i+\frac{1}{2}$ at half of the time interval $t+\Delta t/2$, which is the value at t^n found at $-a\Delta t/2 = -h\nu/2$ away from the interface, or $\frac{1}{2}h(1-\nu)$ from the centre x_i. The average value over the time interval of $\bar{u}^-_{i+\frac{1}{2}}$ to the left of the interface $i+\frac{1}{2}$ is then

$$\bar{u}^-_{i+1} = u_i + \frac{h}{2}(1-\nu)\left.\frac{\partial u}{\partial x}\right|_i = u_i + (1-\nu)\frac{u_{i+1} - u_{i-1}}{4}.$$

The average flux hence becomes

$$\bar{f}_{i+\frac{1}{2}} = f(\bar{u}^-_{i+\frac{1}{2}}) = a\,\bar{u}^-_{i+1} = a\left(u_i + (1-\nu)\frac{u_{i+1} - u_{i-1}}{4}\right).$$

Similarly, at the left interface upwinding uses the linear reconstruction from $i-1$,

$$\bar{f}_{i-\frac{1}{2}} = f(\bar{u}^-_{i-\frac{1}{2}}) = a\left(u_{i-1} + (1-\nu)\frac{u_i - u_{i-2}}{4}\right).$$

Using these fluxes evaluated at time level n in the finite volume method (3.10) one obtains

$$\begin{aligned} u_i^{n+1} &= u_i^n - \frac{\Delta t}{h}(\bar{f}^n_{i+\frac{1}{2}} - \bar{f}^n_{i-\frac{1}{2}}) \\ &= u_i^n - \nu\left(\left(u_i^n + (1-\nu)\frac{u_{i+1}^n - u_{i-1}^n}{4}\right) - \left(u_{i-1}^n + (1-\nu)\frac{u_i^n - u_{i-2}^n}{4}\right)\right) \\ u_i^{n+1} &= (1-\nu)u_i^n + \nu u_{i-1}^n + \frac{\nu(1-\nu)}{4}\left((u_{i+1}^n - u_i^n) - (u_{i-1}^n - u_{i-2}^n)\right) \end{aligned} \tag{4.24}$$

which is called Fromm's scheme. The first two terms on the right-hand side are recognised as the first-order upwind scheme. Fromm's scheme adds to this scheme a correction term that approximates the difference between the gradient approximation around $i+\frac{1}{2}$ and $i-1\frac{1}{2}$. We can observe that the discretisation molecule is not centred around i, but biased upwind: the node $i-2$ enters the difference formula, but not the node $i+2$.

In Sec. 3.1.2 the first-order upwind scheme was applied to the advection of a block profile. Fig. 3.4 on page 48 shows the dissipative behaviour of that scheme which is only mildly improved with mesh refinement.

Fromm's scheme (4.24) produces the results of Fig 4.5, using the same mesh refinement sequence. We can observe the much more rapid improvement with mesh refinement as at each refinement cutting h in half, the truncation error of the advection discretisation is reduced to 1/4. We can also observe that the solution is not monotonic. Small over- and undershoots are created, but the solution is stable: these overshoots do not grow in time but remain bounded.

4.3.4 Monotonicity and second-order accuracy: limiters

The improvement from first- to second-order accuracy with Fromm's scheme is dramatic, but it comes at the cost of losing monotonicity. The exact solution is bounded by the extremal values from the initial solution as $0 \leq u \leq 1$, but Fromm's solution shows small overshoots. These can be problematic, e.g., if the passive scalar u represents the concentration of pollutant that is being advected: what does a concentration of $u = -0.1$ or $u = 1.1$ mean, and how could the numerical method handle that? Or consider the overshoots in a temperature field: the overshooting temperature could erroneously surpass the ignition temperature of a chemical reaction which under correct conditions wouldn't have taken place.

Let us first analyse what effect produces the overshoots. Fig. 4.6 shows the left flank of the block profile advected in Fig. 4.5. Fig. 4.6 (a) shows that the initial solution is bounded and monotonic, but the central reconstruction of the slope used by Fromm's scheme creates an overshoot in the cells either side of the step. The exact advection of this piecewise linear profile is shown in Fig. 4.6 (b). Averaging that profile over the cell produces the new state values for the next time-step which now include an overshoot. Hence the overshoots

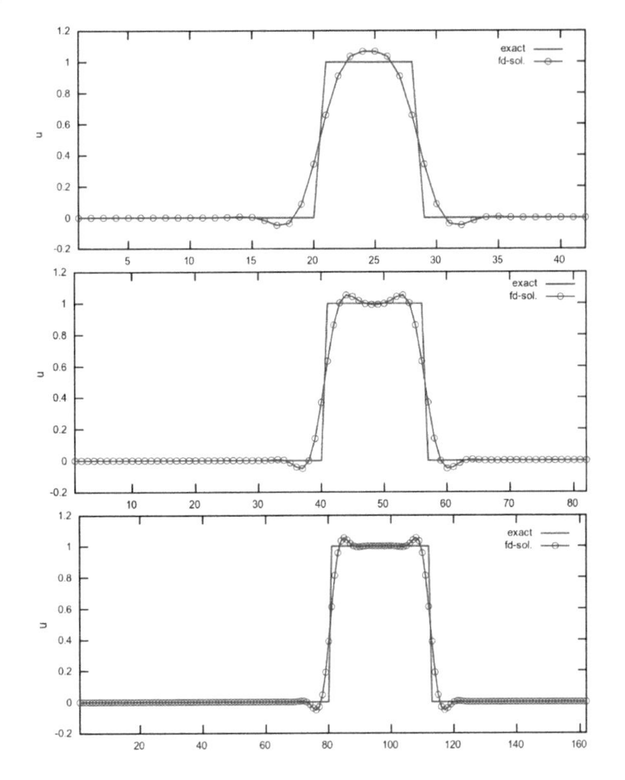

FIGURE 4.5
Advection of a block profile using Fromm's scheme with CFL=0.5. Mesh with 8 (top), 16 (middle) and 32 nodes per block (bottom).

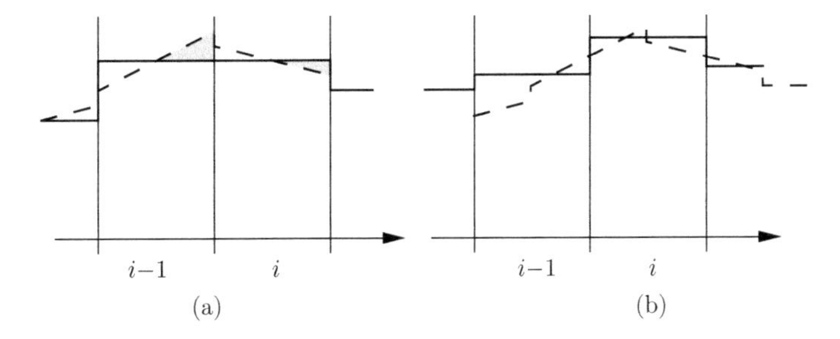

FIGURE 4.6
Step profile (solid line) and its piecewise linear representation (dashed) (a). The flux that varies with time is shown in grey and can be integrated using the trapezoidal rule. (b): advection of the profile with CFL=0.5 (dashed) and averaging for each cell (solid) produces the solution at the new timelevel which exhibits overshoots.

originate from the piecewise linear reconstruction, which is essential to obtain second-order accuracy: both aspects are linked.

So far we have considered discretisation schemes with fixed coefficients, i.e., the discretisation molecule remains the same whatever the solution. It can indeed be proven [1, 2] that it is not possible to have a numerical scheme with fixed coefficients that is second-order accurate and monotonic at the same time. However, there is a loophole. We have so far looked at 'fixed' coefficients as it was simpler, but the finite volume method, of course, is conservative for any consistent discretisation for the flux, as long as it is the same flux seen from either side.

As an alternative we can hence choose e.g., a monotonic first-order upwind scheme where overshoots would be produced, while choosing a higher-order scheme elsewhere. This is the principle of the MUSCL[7] schemes by Van Leer[8] which have been widely adopted in typical CFD solvers. The MUSCL schemes exploit the fact that finite volume schemes using linear reconstruction can be seen as a combination of the basic first-order upwind scheme corrected by contributions from the linear reconstruction. These corrections are directly proportional to the cell gradients used in the reconstruction. Furthermore, as shown in Sec. 4.3.3 we can freely choose the value of the gradient without affecting conservation. Let us reformulate the linear reconstruction of (4.23) as

$$u(x) = u_i + (x - x_i)\,\phi_i \left.\frac{\partial u}{\partial x}\right|_i \quad \text{for} \quad x_{i-\frac{1}{2}} \leq x \leq x_{i+\frac{1}{2}} \tag{4.25}$$

where $0 \leq \phi_i \leq 1$ is a *limiter* function. If $\phi = 1$, we recover Fromm's scheme (4.24) which is second-order accurate but not monotonic. On the other hand, if $\phi = 0$, the state u is taken as constant in the cell; the scheme reverts to first-order upwind which is monotonic. By adjusting ϕ for each cell appropriately, we can ensure monotonicity in critical areas by sacrificing accuracy locally, but maintain overall high accuracy in other regions. Casting this in a slightly more general form and defining the limited slope σ_i as

$$\sigma_i = \phi_i \left.\frac{\partial u}{\partial x}\right|_i \quad with \quad 0 \leq sigma_i \leq \left.\frac{\partial u}{\partial x}\right|_i \tag{4.26}$$

we recover the first-order upwind scheme for $\sigma = 0$, Fromm's scheme for $\sigma_i = \left.\frac{\partial u}{\partial x}\right|_i$.

For one-dimensional analysis as presented here, a number of limiter functions can be derived which each have unique properties that are advantageous in some particular flow situations. The reader is referred to [1, 2, 5] for more detail on this. However, the extension from a scalar advection to systems of equations and from one to three dimensions introduces approximations that impair some of these unique properties. Let us consider here only a basic

[7]MUSCL stands for "monotonic upstream-centred scheme for conservation laws".

[8]Bram Van Leer, Dutch astronomer and mathematician, working in the Netherlands and the USA, contemporary.

example, the minmod limiter, defined as

$$\sigma_i = \text{minmod}\left(\frac{u_i - u_{i-1}}{h}, \frac{u_{i+1} - u_i}{h}\right) \tag{4.27}$$

with the minmod function switching on the minimal magnitude (or modulus) of either argument,

$$\text{minmod}(a, b) = \begin{cases} a & \text{if} \quad |a| < |b| \quad \text{and} \quad ab > 0 \\ b & \text{if} \quad |a| > |b| \quad \text{and} \quad ab > 0 \\ 0 & \text{if} \quad ab < 0. \end{cases} \tag{4.28}$$

We can see that the limiter compares the two slopes $(u_i - u_{i-1})/h$ and $(u_{i+1} - u_i)/h$ either side of i. If the profile of u is linear, both slopes are the same. Using either produces the same result. If the product of the two slopes is negative, triggering the third condition in (4.28), the profile has a maximum or minimum in i. In this case only the first-order scheme produces a monotonic solution; the gradient needs to be limited to zero. In between, the minmod limiter chooses the smaller of the two possible slopes to reconstruct the profile, the most dissipative choice.

Fig 4.7 shows the minmod-limited second-order scheme applied to advection of a block profile. Comparing the limited solution to the unlimited Fromm's scheme is shown in Fig. 4.5.

4.4 Summary of spatial discretisation approaches

In the preceding chapters we have focused on the discretisation of the advective term in the flow equations. Figure 4.8 compares the results of advecting the block profile for two block widths. The differences in the solutions are evident. While the first-order upwind scheme is stable and monotonic, its excessive dissipation makes this scheme less useful for practical application. Switching the space discretisation to a central difference should provide second-order accuracy, but this simplistic modification is unstable. The growing oscillations render this scheme useless. Fromm's scheme, a more sophisticated improvement over the first-order upwind scheme, retains the use of upwinding in the flux calculation, but employs a linear reconstruction of the flux within each cell. The scheme is stable and the accuracy improvements are impressive. However, we lose monotonicity: overshoots arise, but they remain bounded in amplitude.

In most CFD applications we seek an accurate and monotonic flow solution, which can't be achieved with a discretisation that has fixed weights. The remedy to the shortcomings of accuracy of the first-order upwind scheme and of monotonicity for Fromm's scheme is to switch adaptively between the two schemes, depending on the solution. In regions where the gradient changes

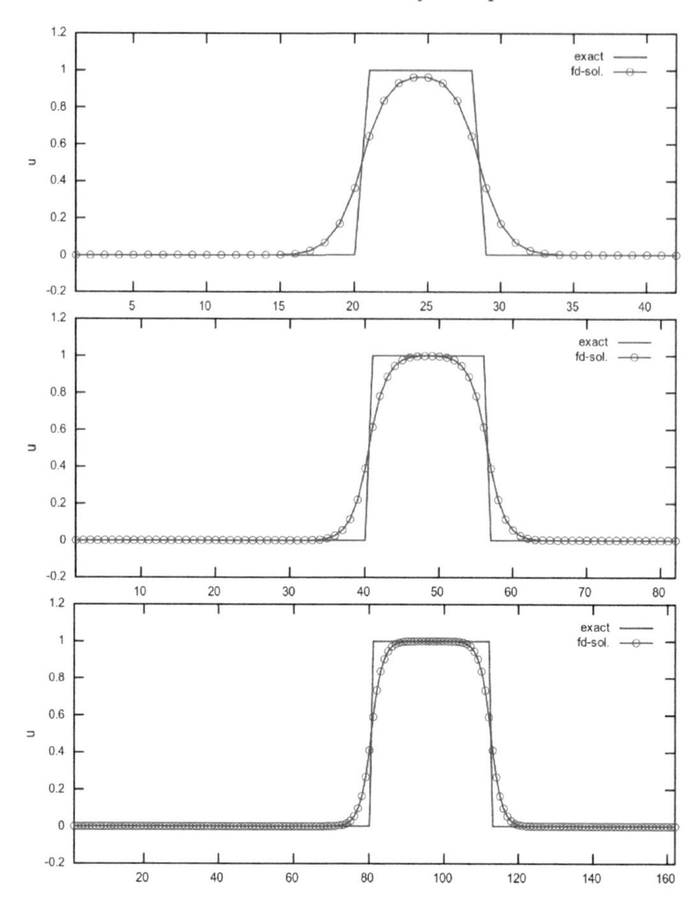

FIGURE 4.7
Advection of a block profile using the minmod-limited linear reconstruction scheme with CFL=0.5. Mesh with 8 (top), 16 (middle) and 32 nodes per block width (bottom).

abruptly or at maxima and minima in the solution, maintaining monotonicity requires using the first-order upwind scheme with piecewise constant data representation. In other regions the second-order scheme with linear reconstruction can be used.

The minmod limiter presented here is the most conservative choice of limiter, using the smallest possible value of the gradient for the linear reconstruction, which results in a scheme that reduces accuracy in many instances, and hence an overall reduction in accuracy. This can be observed when comparing how sharp the block profile is captured between Fromm's scheme and the minmod scheme in Fig. 4.8. Less dissipative limiters can be used, and the numerical solvers, commercial and open-source, typically offer a choice of gradient reconstruction schemes and a choice of limiters.

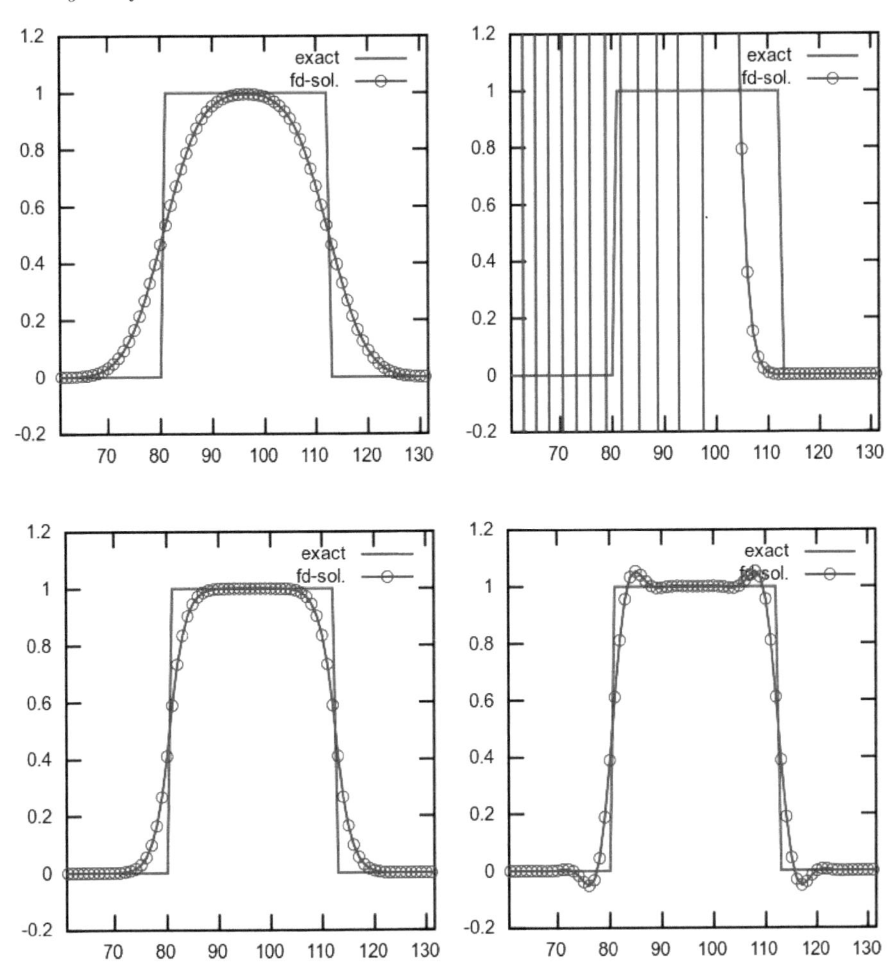

FIGURE 4.8
Advection of a block profile with 32 nodes per block width. Clockwise from top left: first-order upwind, FTCS, Fromm's scheme, minmod limited slopes.

4.5 Convergence of the time-stepping iterations

Section 3.6.3 has introduced the principles of advancing the solution in time through pseudotime-stepping. Section 4.3 discussed the issue of stability and showed how accuracy and stability are competing objectives. This section investigates stability in a more detailed way, considering also how fast a solution is converging, i.e. what the computational cost is.

This section focuses on pseudotime-stepping for the steady state, as in most cases it is this steady solution that is sought after the principal errors of the initial solution have been removed. We use the term pseudotime-stepping since is not important that the solution evolves in a physically accurate way to the steady state: all one cares for is that the *transient* from initial solution to steady state is as rapid as possible with the lowest computational expense. The following sections concentrate on this application.

4.5.1 Explicit methods

Consider again the advection equation (2.35) and its discretisation using the first-order upwind scheme:

$$u_i^{n+1} = u_i^n - \frac{a\Delta t}{h}(u_i^n - u_{i-1}^n). \tag{3.3}$$

The values for the discretisation of the spatial gradient in (3.3) have been chosen from the old timelevel n such that the value at the new timelevel $n+1$ can be calculated explicitly, hence the name *explicit* method. Figure 4.9 shows the *molecule* of the points which are involved in this discretisation. The factor $\nu = \frac{a\Delta t}{h}$, the *Courant number*, has been introduced in Sec. 3.1.1. The terms on the right side can be regrouped as

$$u_i^{n+1} = (1-\nu)u_i^n + \nu\, u_{i-1}^n. \tag{3.4}$$

This shows that the first-order upwind scheme actually uses a weighted average of the values at i and $i-1$ for the update at i, provided $0 \leq \nu \leq 1$. An average means that the value of the average is bounded by the minimum and maximum of the values to be averaged. In (3.4) the weights for u_i^n and u_{i-1}^n are positive and u_i^{n+1} cannot be larger or smaller than the values at the old time-step; the discretisation is monotonic (see Fig. 3.2 in Sec. 3.1.4).

The condition that

$$0 \leq \nu \leq 1$$

is actually also the limit of stability for the first-order upwind scheme 3.4: the *Courant number* ν has to be positive but less than one. This condition is also known as the *CFL condition* (cf. Sec. 3.1.1).

The condition that $\nu > 0$ results from the fact that the backward difference is only upwind if $a > 0$, for the reverse case of advection from the right, $a < 0$ a forward difference would have to be used in an upwind scheme. Both can be combined when using this form of first-order differencing,

$$\left.\frac{\partial u}{\partial x}\right|_{upw} = \frac{|a|-a}{2h|a|}u_{i+1} + \frac{a}{h|a|}u_i - \frac{|a|+a}{2h|a|}u_{i-1} \tag{4.29}$$

It is left to the reader to show that this works for both directions in Exercise 4.5 at the end of the chapter. For simplicity, let us consider in the remainder only the upper limit of the CFL condition, $|\nu| < 1$.

If the CFL condition is not satisfied, Eq. (3.4) is no longer an averaging formula. The negative coefficients make it unstable: a numerical example — or

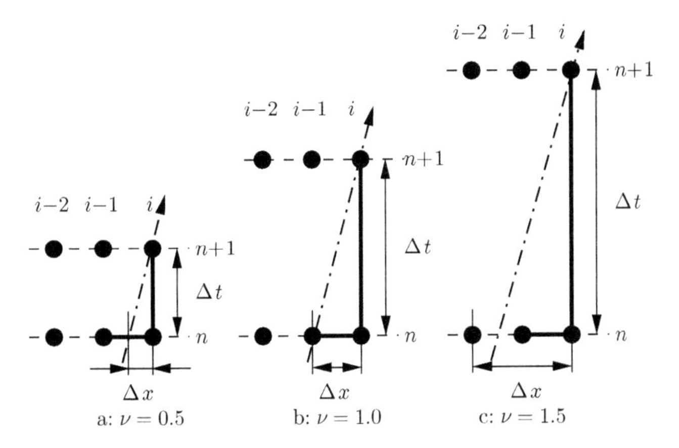

FIGURE 4.9
Molecule (solid lines) of the first-order upwind scheme. Timelevels n and $n+1$ are dashed; the advection speed (dash-dotted vector) is constant; the time-step is varied to obtain $\nu = 0.5, 1.0, 1.5$. For $\nu = 1$ the solution moves to the right by exactly one grid point. For $\nu = 1.5$ the discretisation molecule does not extend far enough back in space to include the necessary information.

analytical stability analysis — shows that oscillations arise at steep gradients and are amplified for $\nu > 1$. The *stability limit* for the first-order upwind scheme is $\nu \leq 1$.

It is instructive to consider what the limiting value of $\nu = 1$ corresponds to. Since $\nu = \frac{a\Delta t}{h} = 1$, $h = a\Delta t$, that is, the information that is carried with speed a crosses exactly one cell of width h in one time-step of Δt at $\nu = 1$. Fig. 4.9(b) demonstrates this. In this case the first-order upwind scheme is actually exact: all errors vanish. However, a uniform Courant number of $\nu = 1$ is actually not achievable for the Navier-Stokes equations which have a number of transport speeds, also called *characteristic speeds* as discussed in Sec. 5.2.7.1.

In the case $\nu > 1$ information from $i-2$ is required, but the molecule of (3.4) contains only the first point to the left, $i-1$. In this case not all relevant information is taken into account by the scheme and it cannot be stable, as shown in Fig. 4.9(c). Similarly $\nu < 0$ would require the point $i+1$ to be present in the molecule, which we can include with the modification (4.29).

4.5.2 Implicit methods

When deriving the first-order upwind discretisation (3.3), the spatial discretisation was taken at the old, known timelevel. Alternatively, one could choose to approximate the space derivative at timelevel $n+1$ (for simplicity limiting

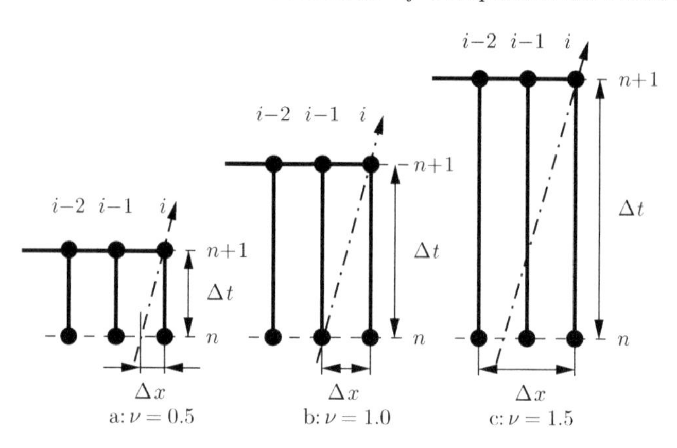

FIGURE 4.10
Implicit molecule.

the validity to $a>0$):

$$\frac{u_i^{n+1}-u_i^n}{\Delta t} = a\frac{u_i^{n+1}-u_{i-1}^{n+1}}{h}$$
$$(1-\nu)u_i^{n+1}+\nu\, u_{i-1}^{n+1} = u_i^n. \tag{4.30}$$

The molecule is shown in Fig. 4.10(b). Note how the molecule is just mirrored in the time-direction by switching n and $n+1$, as are the discretisations in Eqs. (3.4) and 4.30.

How can we solve this equation? We find two unknowns, u_{i-1}^{n+1} and u_i^{n+1}, but only a single equation. We could recruit the shifted equation for $i-1$, which in turn adds unknown u_{i-2}^{n+1}, and so on. The chain only stops when we hit the boundary where the value is specified through a boundary condition. Fig. 4.10 shows the molecule.

There is a significant advantage to this discretisation. Since all the nodes along x are involved in the molecule we always have the correct values u^n included in the molecule; the method is unconditionally stable. Very large values of the Courant number CFL can be chosen to accelerate the convergence to a steady state, without risking amplification of error modes — the formal analysis is left to e.g. Hirsch [2]. This is an important advantage especially when the equations are very tightly coupled as is the case with turbulence models.

However, it is now no longer possible to explicitly solve for the updated state u^{n+1}. The values at the new time-step are defined by a coupled system of equations of the form $\mathbf{A}x = b$, whose solution requires a significant numerical effort. In practical calculations the systems are too big to be solved directly with direct Gaussian elimination techniques but the solution to this system of equations needs to be approximated iteratively. Most iterative methods that are used typically require the storage of the system matrix A which has

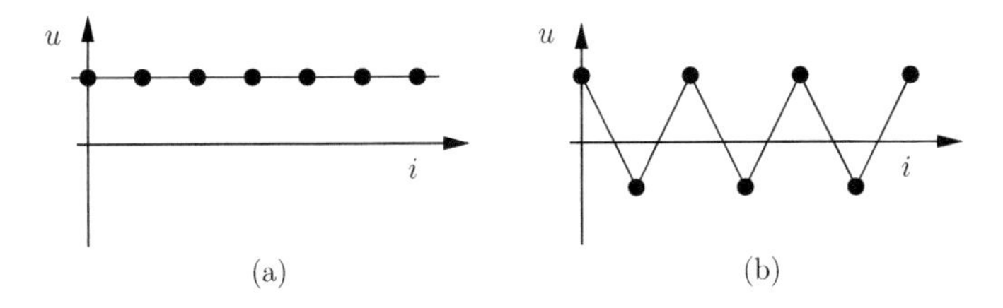

FIGURE 4.11
Lowest (a) and highest (b) frequencies on a mesh.

dimension $N \times N$ for a grid with N nodes. The matrix will have non-zero entries only where neighbouring cells interact through a flux interface, and storage can be reduced exploiting that sparsity, but there is still a significant storage requirement.

4.5.3 Increasing mesh resolution

In Section 4.3 it has been shown that stability requires that no modes are amplified. Now is the time to explain a bit more what "mode" refers to. In order to investigate stability the amplification of a given input mode is measured; for a stable scheme this amplification factor G has to be less than one, $G \leq 1$. We can conveniently express the variation of the solution as a sum of Fourier modes, $u = \hat{u} cos(if)$, where $\hat{u}$ is the amplitude and i is the spatial index of the grid point. We can then consider the behaviour of modes with different *frequency* f.

A constant, uniform field is of the lowest frequency one can consider (Fig. 4.11(a)). A constant is not oscillating at all; it has frequency $f = 0$. The highest frequency that can be expressed on a mesh is a *sawtooth* mode (Fig. 4.11(b)). The mode has one wavelength spread over two mesh widths h; hence the highest possible frequency on a grid is $f = \pi$.

These frequencies see different amplifications in one time-step. The lowest frequencies correspond to a constant, which must be preserved in a consistent, stable scheme, and hence their amplification factor must be $G = 1$. No frequencies may be amplified, $G \leq 1$ for all frequencies. It is not possible to devise a discretisation that maintains $G = 1$ for all of them; hence, in practice all of the higher frequencies are damped, $G < 1$. A typical amplification graph is shown in Fig, 4.12. This damping comes about due to artificial viscosity in the discretisation.

It can be observed that the highest frequencies are damped the most. A typical first-order discretisation with an artificial viscosity proportional to the mesh width h will damp high frequencies much more than a second-order method with artificial viscosity proportional to h^2.

In most instances high frequencies are errors which are still present from the initial solution. The artificial viscosity actually has a welcome effect on

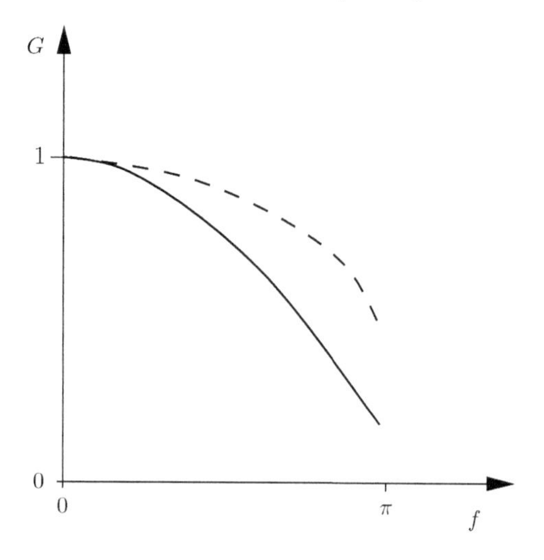

FIGURE 4.12
Amplification of a range of frequencies. A typical first-order accurate scheme is shown with the solid line, a typical second-order one with the dashed line. Note that the amplification factor G will strongly depend on the CFL number and the iterative scheme (time-stepping).

these errors: they are strongly removed by damping. The better the damping, the faster the convergence to the steady state where all of these errors have been removed. High frequencies and first-order methods have better damping.

A well-discretised solution will have many points to represent a particular feature, e.g. 15-30 points across a boundary layer profile, it will have mostly low frequencies. Ideally, however, we would like to avoid fine meshes and would like to be able to represent high frequencies accurately as well. Also there are instances where resolution of a feature with few mesh points cannot be avoided, resulting in high-frequency modes which are not errors. Examples are shocks which are always captured over 2-3 mesh points or flame fronts. These feature modes are smeared in the same way as error modes by the artificial viscosity.

In summary, this has two effects on the solution with respect to accuracy and convergence:

1. The first-order solution produces more artificial viscosity and has less accuracy especially for the high frequencies. This is a problem where the high frequencies are features of the solution.
2. All the high frequencies which are errors are damped much better by the first-order solver. It shows a much better convergence rate.

Conversely, second-order accurate methods provide better accuracy: features need to be resolved over fewer mesh points to obtain the same accuracy. On the other hand, second-order methods converge much more slowly.

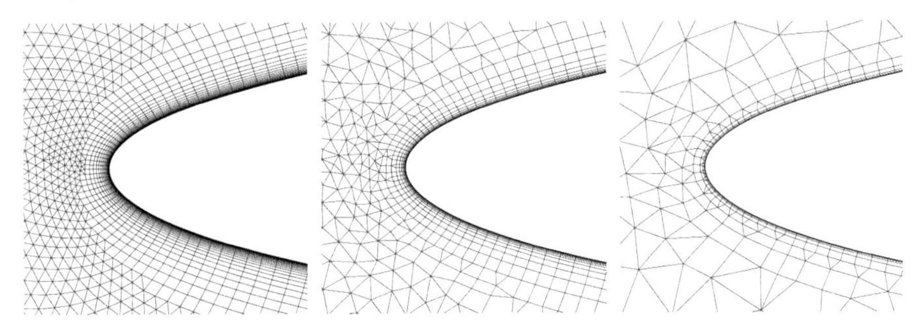

FIGURE 4.13
Coarsening a hybrid grid around a RAE aerofoil. The finest and two coarser levels are shown around the leading edge.

4.5.4 Multigrid

Section 4.5.2 has discussed the advantages of implicit methods in converging much more quickly to a steady state in Section 4.5.2. This comes at a cost of very high storage. An alternative method to obtain a temporal discretisation with faster convergence and better stability properties is to use a number of grids covering the same domain, one coarser than the other. Fig. 4.13 shows an example of a fine grid and two coarsened levels for an aerofoil mesh.

Multigrid methods in CFD typically use a sequence of 3-5 meshes, usually coarsened by a factor of 2 in each direction. The solver then cycles between the meshes and uses updates from the coarser meshes to advance the solution on the finest mesh. Multigrid methods can accelerate the convergence dramatically. A convergence rate of 10 times more is not unusual. Fluent offers a multigrid option for the compressible code.

Using a sequence of coarser grids helps to accelerate the convergence in two ways:

1. The stability limit increases on coarser grids due to the larger cells; the solution can be run with larger time-steps. Errors traverse the domain and exit through the boundary much more quickly.
2. A grid which is twice as coarse uses half the number of points to resolve the same feature. Hence, the frequency is doubled. For example, consider a sine wave represented in 4 points per wavelength, frequency $f = \frac{\pi}{2}$. On a coarser mesh the same wave is resolved in only two points per wavelength, resulting in twice the frequency $f = \pi$, the highest frequency which can be expressed on a mesh. Error modes of that frequency see the strongest damping; hence, the errors are damped out more rapidly.

4.6 Excercises

4.1 Apply Taylor analysis to show that the central discretisation 3.16 of the second derivative is second-order accurate.

4.2 Apply Taylor analysis to Fromm's scheme, Sec. 4.3.3, and analyse its order of accuracy.

4.3 Implement Fromm's scheme in a spreadsheet using the conditions of Fig. 3.3. Run advection simulations with smooth profiles (sine curves, smooth variation) and simulations with block profiles (steep gradients).

4.4 Implement the minmod-limited linear reconstruction scheme in a spreadsheet using the conditions of Fig. 3.3.

4.5 Show that the two-sided upwind difference (4.29) reverts to the backward difference for $a>0$ and the forward difference for $a<0$.

5

Boundary conditions and flow physics

Boundary conditions (BCs) allow the user to choose the flow conditions once the domain of the simulation is fixed. Clearly, only the assignment of specific boundary conditions make a simulation unique and applicable to the problem that is being simulated. The correct selection of BCs for a particular boundary needs to consider the combination of all BCs around the domain. Sec 5.3 will have a more mathematical look at what combinations can work and which ones can't. This requires the understanding of how information travels in the flow field which is discusssed in Sec. 5.2. Let us start with some considerations based on the flow physics to get started.

5.1 Selection of boundary conditions

5.1.1 Some simple examples

In some cases the assignment of boundary conditions is very straightforward. Consider e.g., the flow near a solid wall, the *wall condition*. Depending on what type of mathematical model is being used, and what behaviour we want to impose there, we need to choose an appropriate condition.

In a viscous Navier-Stokes model a *no-slip* condition is imposed on a fixed wall; the flow has zero velocity at the wall.[1] If the wall is moving, e.g., a rotating tyre, the no-slip condition means that the the velocity is zero with respect to the moving surface, i.e., the fluid at the wall moves with the rotational speed of the tyre.

If an Euler model is used for the flow where viscous effects are neglected, a *slip-wall* condition needs to be chosen which imposes *tangency* of the flow at the wall: the velocity at the wall has to be parallel to the wall, but cannot penetrate it. It may also be appropriate to choose a slip-wall condition in viscous flow, e.g., when simulating a vehicle in a wind-tunnel where the viscous effects on the tunnel wall are not of interest and we don't want to refine the grid there to represent these effects.

For inflow and exit boundaries, the conditions have to be considered more carefully and in conjunction.

[1] To make the simulation simpler, aerofoil or aircraft simulations are done similar to wind tunnel experiments: the aircraft is fixed, and the flow is blowing onto it.

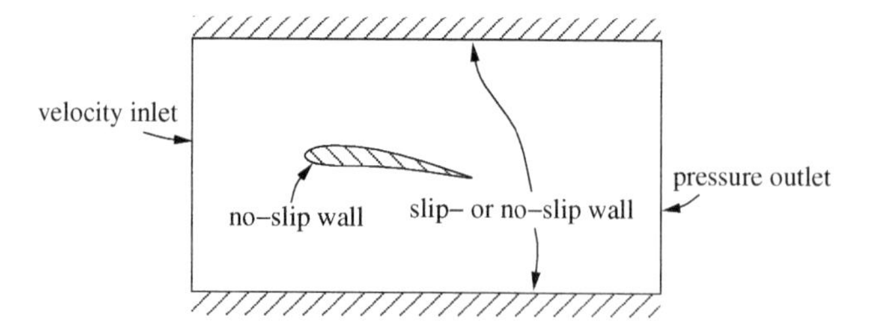

FIGURE 5.1
Boundary condition setup for viscous incompressible flow in a channel or wind tunnel with vane. If the boundary layer at the top and bottom walls of the wind tunnel are not of interest, a slip-wall condition may be appropriate also for this viscous flow.

A typical combination of boundary conditions (BCs) e.g., for incompressible flow in a channel, would be to impose the velocity vector (e.g., its components) at the inlet and the pressure at the outlet as shown in Fig. 5.1.

The same strategy when applied to the farfield boundary of an aerofoil in free flight would require that the part facing the flow (and maybe some of the exit part) receives the velocity inlet condition, while at least some of the exit section needs to be specified with the pressure outlet condition. Fig. 5.2 shows a suitable choice of portion of the outflow section where the pressure condition is assigned. It is simpler to rotate the incoming flow rather than the aerofoil geometry to simulate a change in angle of attack. By restricting the pressure condition to a smaller rear-facing angle we ensure that the flow always exits in that portion of the boundary, even at higher angle of attack.

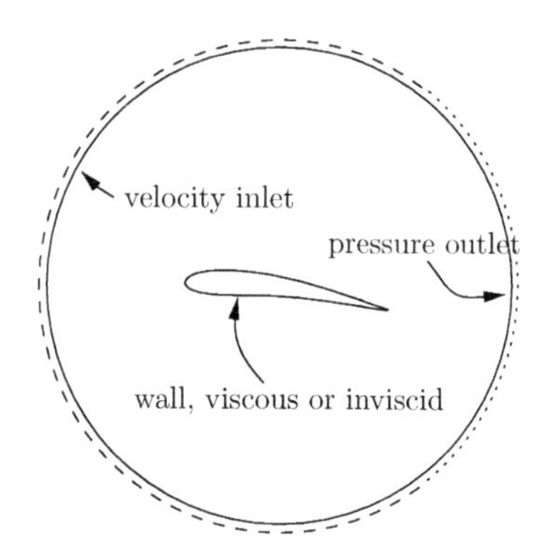

FIGURE 5.2
Boundary condition setup for aerofoil in incompressible flow; the pressure outlet condition (dotted) will experience only exiting flow under moderate changes of angle of attack imposed by changing the direction of the incoming flow of the velocity inlet condition (dashed).

Let us consider how continuity is satisfied with this choice of boundary conditions (BCs). At the inlet boundary the incoming mass flow is fixed as the BC specifies the velocity profile and density is constant. The finite volume method for the continuity equation is conservative; hence, it is guaranteed that inside the domain fluxes cancel each other as what leaves one cell is exactly the same as what enters the neighbouring cell. Hence by considering a control volume around the domain it is clear that the integral of the mass flow at the exit is fully determined by the integral of the mass flow at the inlet: it has to be conserved.

This in turn explains that we actually must not fix the velocity on at least some part of the exit; otherwise, the exit velocity profile could not adjust to satisfy continuity: if we were to prescribe more flow going in than we prescribe to exit, the mass in the domain would permanently increase.

Note that even if we knew the value of velocity at the exit and imposed that, we could only impose that to the finite precision that numbers are stored within the computer, e.g., $1 \cdot 10^{-7}$ for single-precision arithmetic with seven digits, similar to your pocket calculator (cf. Sec. 4.2.1). Over a large number of iterations even such small differences would add up and corrupt the simulation.

Similarly we can consider how the momentum equation is satisfied. Recall from Sec. 2.3 that the momentum equations describe the balance of momentum in each direction: the difference between the momentum in and out of the control volume is equal to the sum of the forces. The finite volume method again conserves the momentum inside the domain, so we can again consider the entire domain as a control volume. The momentum transport in and out of the domain is fixed at the inflow by the inlet BC and at the outlet by continuity.

The resulting difference in momentum has to be equal to the sum of forces. Viscous forces at the wall depend on viscosity and the velocity gradient, hence are also determined by the velocity field. The main forces that remain to be considered are, hence, the pressure forces on the walls and on inlet and outlet. At the outlet the pressure outlet condition fixes the pressure, but we have not set a boundary condition for the pressure at the inlet or at the solid walls. These pressures must be left free to adjust in order to satisfy the momentum equation. We are not allowed to specify the inlet pressure for this flow if we specify outlet pressure and inlet velocity.

5.1.2 Selecting boundary conditions to satisfy the equations

While the combination of inlet velocity and outlet pressure boundary conditions is a very common one, many other combinations are possible, e.g., we could assign inlet pressure as well as outlet pressure and have the velocity adjust to satisfy the momentum equation. However, as Sec. 5.2 will explain, the inlet pressure condition then needs to impose more than just the pressure.

Some combinations of boundary conditions, however, cannot satisfy the flow equations. Fig. 5.3 shows the domain for a blood vessel bifurcation. We

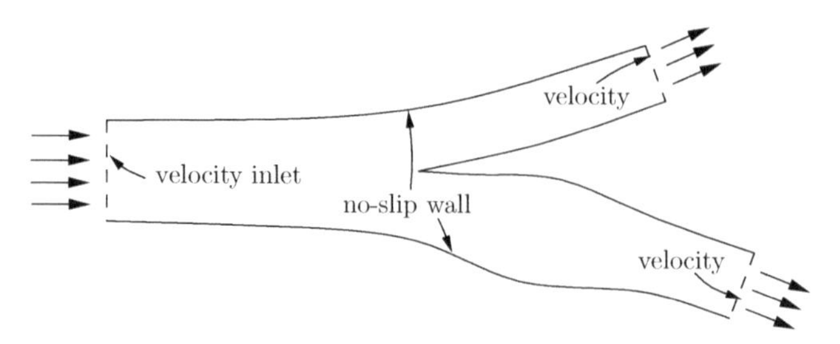

FIGURE 5.3
Simplified carotid bifurcation.

want to impose the overall flowrate going in, as well as the split in flowrate to the two daughter vessels, hence the proposal to fix the incoming flowrate and then the split through two velocity conditions in the exit sections.

This combination will not lead to a solution. The velocity prescribed at all inlet and outlet sections is over-specified. There is no boundary condition where the exiting velocity can adjust to satisfy continuity. The pressure on the other hand is underspecified: it is not fixed anywhere. Recall that the absolute pressure does not appear in the incompressible Navier-Stokes equation; only the pressure gradient does. The pressure difference between inlet and outlets is needed to close the momentum balance. For this a reference value is needed relative to which the inlet and outlet pressures can be defined.

To fix this configuration, we could reassign one of the outlets to have a pressure outlet condition and impose e.g., a gauge pressure of zero there. The mass flow exiting at that BC by conservation has to be the difference between the imposed inlet mass flow and the mass flow drawn off at the other exit section. The pressures in those sections can then adjust to satisfy the momentum balance.

5.2 Characterisation of partial differential equations

At the heart of the assignment of boundary conditions is the analysis of how information travels in a flow. While the physics of the flow equations can give some insight as discussed in the preceding chapter, we need to understand the mathematical character of the equations since that determines how the flow properties or any perturbations in the flow field are transported by the flow across cell interfaces and into and out of the domain. As we have seen from the example of the discretisation of the advection equation in Sec. 3.1, for this type of equation incoming information needs to be imposed; exiting information comes from the solution inside the domain. However, the analysis is no longer that simple when we consider coupled systems such as the flow equations.

The formal task in selecting a boundary condition is therefore to identify in the equations how information travels. The user needs to understand which information enters and which leaves the domain, what types of boundary conditions this allows and how each condition acts within the 'concert'[2] of the other boundary conditions applied.

The Navier-Stokes equations apply equally to liquids and to gases — provided one chooses a suitable equation of state to close the system. Hence, one will expect CFD to apply to fluids in general, as well, and this is the case. However, the character of the equations changes as it reproduces the changing physical characteristics in different flow regimes, e.g., supersonic flow over an aircraft behaves very differently from, say, the flow in the heart. The numerical method also has to reproduce this behaviour properly. The discretisation has to be chosen suitably and, most importantly, boundary conditions need to match the physical flow of information.

In the following a few very simple problems are presented, each which embodies a characteristic type. The final section will briefly consider what character the full Navier-Stokes equation takeon in different regimes.

5.2.1 Wave-like solutions: hyperbolic equations

The advection equation describes pure advection of information at the constant speed a. If the initial solution at $t = 0$ is a pulse as shown in Fig. 2.4, the pulse will be translated by $x = at$ at some later t. Problems like this are also known as marching problems, since one can obtain a solution by simulating the evolution of the flow over time, or "marching" it in time.

In this simple equation information only travels from left to right in the positive x-direction, provided $a > 0$. This behaviour is called *hyperbolic*. In hyperbolic equations *characteristics* can be identified which are the trajectories along which a certain quantity, an *invariant*, is constant. In the linear advection equation, the passive scalar u is constant along lines of $z = x + at$. We could also interpret the changes in the solution resulting from the addition of a large number of small steps, each resulting from a wave that brings a step change and travels with a particular speed.

In the case of scalar advection, equation 2.35, the choice of boundary condition is very simple: a value needs to be specified at the inlet on the left; no value may be specified at the outlet on the right.

[2]'Concert' is a very appropriate image here: the methods we use to solve the discretised system of equations are iterative, which can be interpreted as some form of time-stepping with locally dilated times. But the underlying physics of conservation built into the method remain: during the convergence waves of perturbations or errors produced by the 'bang' of starting the flow with an initial but incorrect solution will bounce back and forth through the domain until all the errors are either dissipated or have left the domain. The way the errors reflect on 'echo' between boundary conditions has a major effect on the convergence.

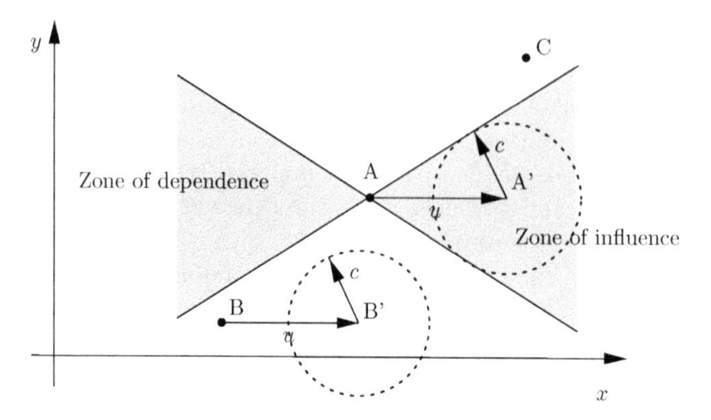

FIGURE 5.4
Zones of influence and dependence in 2-D supersonic flow. The circle indicates the distance a sound signal has travelled in the time interval for the flow to transport from A,B to A',B'.

Another example of hyperbolic character is steady two-dimensional supersonic flow. In this case there is no longer a single equation, but four equations for density; x- and y-momentum and energy are coupled together. Information in a fluid can propagate at most with the speed of sound: if the fluid itself is travelling faster than the speed of sound, information cannot travel back upstream. In a supersonic field, given some location x, y, a zone of influence can be identified downstream of where a perturbation such as e.g., an aircraft is placed; see Fig. 5.4. In three dimensions this zone will be a cone, the Mach[3] cone. Similarly upstream of a point is its zone of dependence. Only perturbations in the flowfield inside that zone can reach the point; any perturbations outside it will have no influence.

In Fig. 5.4 any information from point A can only reach locations in its zone of influence downstream. Point C can never receive any information from A; neither can B influence A since B is not located within A's zone of dependence.

This physical behaviour, or mathematically speaking the hyperbolic character of the equations, has a bearing on the selection of a discretisation. Clearly, a discretisation for the point A which does not take into account all of the information in its zone of dependence will produce significant errors or even be numerically unstable. When working with commercial flow codes all of this has been taken care of. We have respected the domain of dependence when limiting the Courant number of our explicit discretisation of the advection equation to $\nu \leq 1$ in Sec. 4.5.1.

Supersonic flow fields can exhibit shocks, discontinuities in the flow states. More generally speaking, hyperbolic systems allow discontinuous solutions.

[3] Ernst Mach, Austrian physicist and philosopher, 1838-1916.

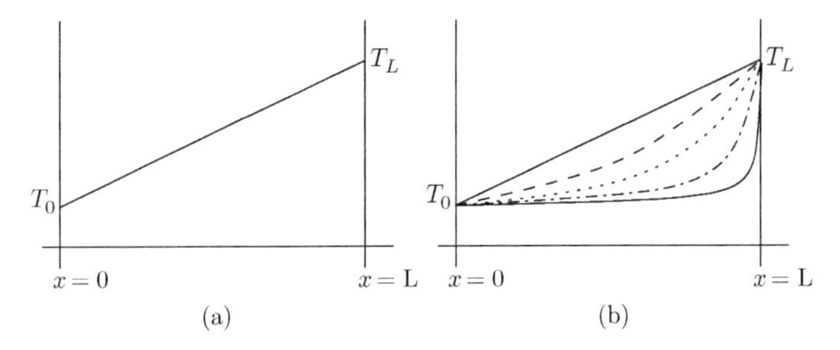

FIGURE 5.5
Heat conduction in a rod, (a): steady, (b): unsteady.

5.2.2 Smoothing-type solutions: elliptic equations

Another major class of problems is often called equilibrium problems. The type of the equations governing these problems is called *elliptic*, e.g., consider steady state heat conduction in a rod held at different temperatures at either end $x = 0$ and $x = L$, as shown in Fig. 5.5. The heat equation has been derived in Sec. 2.4.3:

$$\frac{\partial^2 T}{\partial x^2} = 0. \tag{2.44}$$

The solution is a linear temperature proflie, Fig. 5.5 (a).

The steady state solution does not depend on the initial temperature distribution in the rod. It only depends on the values of the temperature at either end of the rod T_0 and T_L. Changing either value will change the temperatures in all points along the rod. For a numerical scheme this means that information from all directions has to be taken into account and that a discretisation method needs to use information from all neighbouring points. For the boundary conditions this means that a condition needs to be specified at all boundaries. We could impose the fixed values of temperature as in Fig. 5.5, but we could also impose the heat flux, i.e., the gradient of the temperature. For example, zero heat flux or zero gradient corresponds to an adiabatic or fully insulated end.

It can be shown that solutions to elliptic problems are smooth. Discontinuities can be imposed at boundary conditions, but the elliptic behaviour will smoothen it out in the interior of the domain.

5.2.3 The borderline case — parabolic equations

A better understanding of elliptic problems can be gained when considering the time-evolution of the heat equation. The unsteady heat equation was derived in Sec. 2.4.3:

$$\frac{\partial T}{\partial t} + k\frac{\partial^2 T}{\partial x^2} = 0 \tag{2.43}$$

where k is the thermal conductivity. Figure 5.5(b) shows a rod which at time $t_0 \leq 0$ is at temperature $T = T_0$. At $t = 0$ a Bunsen burner is held on the end at $x = L$ which instantaneously raises the temperature at that end to $T = T_L$. In the steady state one ends up with the elliptic problem of Eq. 2.44, which is independent of the initial condition. The transient to that state, however, will depend on the initial state. The problem now has become a marching problem — the solution can be advanced in time.

Mathematically, the "sound speed" at which information travels in parabolic equation is infinite: the point at $x = 0$ immediately "knows" about the flame at the other end, but the limits of heat diffusion only allow very little energy to be transferred, changing the temperature profile very gradually, but everywhere. This is called a *parabolic* problem. As opposed to the hyperbolic case, a characteristic speed with which information travels cannot be identified in this problem; hence, there are no characteristics in parabolic equation.

As opposed to the elliptic problem the parabolic case does know a "time-like" direction. Everything in the past is known — the initial state influences the transient at every location — and every location in the future is influenced. Hence, we can consider the parabolic case as the limiting case between hyperbolic and elliptic behaviour. Different though from the hyperbolic case, the parabolic character does not have characteristics: we cannot identify particular directions of waves that carry particular information.

5.2.4 The domain of dependence, the domain of influence

Figure 5.6 sketches the domains of dependence and influence, for example, of flows exhibiting each of the three characters. In the hyperbolic case the flow state at a point depends on all values in the Mach cone upstream of it, the shaded area in (a). The sound speed measures how fast information travels, $c = \Delta x / \Delta t$, which is the inverse of the slope of the Mach cone. The two limiting slopes of the Mach cone are characteristic directions, e.g., a shock wave in supersonic flow is the fastest wave and would propagate along this line.

Increasing the speed of sound, or, conversely, reducing the speed of the flow, flattens the Mach cone. In the special case where flow speed u and sound speed c are equal, i.e., the Mach number $\text{Ma} = u/c = 1$, the slope of the cone is horizontal. This is the parabolic case shown in Fig. 5.6(b): the flow at a given point depends on the flow at all points in the past and all points of the present, i.e., the same time level. Perturbations to the flow at a given point influence all of the future. While the time still has the meaning of separating past and future, we cannot formulate characteristic directions along which information propagates.

Reducing the Mach number to subsonic speeds, $\text{Ma} < 1$, i.e., making the sound speed larger than the flow speed, allows perturbations to reach any part of the domain. This is the elliptic case, (c) in Fig. 5.6. Characteristic

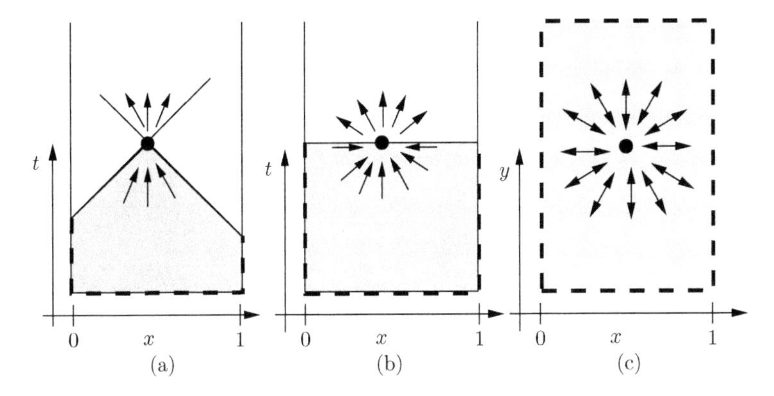

FIGURE 5.6
Domains of dependence (shaded) for (a) hyperbolic, (b) parabolic and (c) elliptic equations. Arrows indicate how information can travel; the dashed line indicates where initial or boundary conditions need to be imposed.

directions of propagation cannot be defined. All points depend on the entire domain; all points influence the entire domain.

5.2.5 Example of characterisation: surface waves

As a practical example of this behaviour we could consider surface waves on water; hence, the transported variable is water height. The finite speed[4] of the wave is the speed at which any height changes are transported through the domain, in other words "communicated" to other regions of the flow field.

Let us first consider the "supersonic" case when the water travels faster than the speed of the surface waves. Figure 5.7 shows a water channel that was meant to supply the water wheel of an old sawmill. In this type of "shallow water flow" the velocity and height of water in the channel are coupled through the momentum and continuity equations.

In this particular example the flow is running at about twice the speed of the surface waves. The sudden change in direction at the left corner of the channel creates a step change in height. This perturbation propagates away from the corner with the speed of the surface wave in all directions, but by the time the surface wave has travelled one channel width upstream, it has been transported downstream by two channel widths. Hence information cannot travel upstream; the incoming flow remains unaware of the kink until the flow hits it. Note that giving the wave more time to reach upstream does not help; we are looking at the steady state, so already are considering infinitely long times.

[4]Surface waves are actually *dispersive*, i.e., the wave speed depends on the height of the wave; taller waves travel faster. For the small changes in height we consider here, we can take this speed as constant.

FIGURE 5.7
Surface wave on a water channel supplying a mill. The water enters at the bottom right, running faster than the speed of the surface waves. The change in direction creates a surface wave that is reflected at the channel sides. Information about the change in direction or any changes in water height cannot propagate upstream.

Similarly, by the time the perturbation has spread across the channel, the water has transported that perturbation downstream twice as far, which explains the angle of the wave. The triangle of flow to the right of the first wave cannot receive any information from the corner; it is not in its domain of influence. This flow situation represents the hyperbolic case (a) of Fig. 5.6.

Let us now consider the alternative case where the flow is slowed down by e.g., closing the inlet shutter to the channel such that the surface wave speed is faster than the flow speed. The waves still travel at (roughly) the same speed as before, but given enough time they can reach any location in the channel. And if we consider the steay case, this means we consider infinitely long times. This flow situation represents the elliptic case (c) of Fig. 5.6.

In between these two states is the parabolic case, where the flow speed in the channel is exactly equal to the wave speed. We could reach this state starting from the high-speed state in Fig. 5.7 and gradually reduce the flow speed. The angle between the surface wave and the flow direction (or the channel wall), the *Mach angle*, would become gradually larger and larger until in the case of equal wave and flow speeds, the wave is perpendicular to the flow direction. This flow situation represents the parabolic case (b) of Fig. 5.6. Information still can't travel upstream since the perturbation that tries to propagate upstream with the wave speed is transported downstream equally fast with the flow speed. Hence the incoming flow cannot know about the jump in water height until it hits it. But all the points along that wave, i.e., across the channel's width, are aware of each other as the perturbations can travel across the channel.

5.2.6 Compressible and incompressible flows

In practice it is often simpler to neglect effects of compressibility if the flow speed is low enough to allow this, and if other compressible effects such as sound waves are not relevant. This in turn implies changes in the discretisation, considered in Sec. 3.6.6, and consequently in imposing boundary conditions, considered in Sec. 5.3.1.

A good understanding of the differences between compressible and incompressible flow can be gained by starting from the surface wave model of Sec. 5.2.5 and replacing the surface waves with acoustic waves. Acoustic waves, or sound waves, are small fluctuations in pressure that we can hear. Changes in the pressure field can be built up of a superposition of waves; similarly, changes in the pressure field will generate acoustic waves. Sound waves propagate with the speed of sound which depends on density and temperature for an ideal gas, but which for simplicity here we can assume to have a constant speed.

Let us first consider the low-speed case, and to begin with the unsteady situation, e.g., a speaker in a lecture hall. The voice will generate acoustic waves which travel at a finite speed c. In a given short time interval Δt the waves can only travel a finite distance of $\Delta r = c\Delta t$. The domain of influence is the radius Δr. We are observing in time; there is a time-like direction. The rays from the speaker into the lecture hall are characteristic directions along which the acoustic information travels. In the surface wave model of Sec. 5.2.5 this would correspond to throwing a stone in a pond and watching how the waves ripple out and are reflected back. This problem is by nature compressible, as a small pressure fluctuation corresponds to a small density fluctuation. Both types of fluctuation are coupled through the ideal gas equation. Density cannot be constant. The unsteady, compressible Navier-Stokes equations, hence, are hyperbolic.

What if the flow in the lecture hall is stagnant, after we closed doors and windows? When claiming that low-speed flow can be considered incompressible, this only applies to the steady state or to time-scales of observation that are much, much longer than the time it takes for an acoustic wave to traverse the domain. Taking this point of view means to consider the long-term behaviour once all acoustic waves have dissipated. In the surface wave model of Sec. 5.2.5 this would correspond to throwing a stone in a pond and coming back the next day to see how much the water level has risen, which implied to assume that there are no surface waves and that the level of the water table is the same across the pond. The equivalent in incompressible flow is to assume that density is constant and there are no acoustic waves.

5.2.7 Characterisation of the Navier-Stokes equations

5.2.7.1 The compressible flow equations

As discussed in the preceding section, the unsteady compressible Navier-Stokes equations are always fully hyperbolic. This implies that we can view any small change in the solution as being brought about by a number of waves which travel in a particular direction, the *characteristic direction* with a particular speed, the *characteristic speed* and bring about a particular type of change.

Without entering into the mathematics of determining the characteristics and their speeds, see e.g., [2] for details, the two-dimensional compressible flow has the following characteristics:

- forward-running acoustic wave, speed $u + c$,
- enthalpy wave, speed u,
- entropy wave, speed u,
- backward-running acoustic wave, speed $u - c$.

We can, hence, exploit this hyperbolic character to design numerical schemes and to select boundary conditions.

Hyperbolicity means that we can always define characteristics. In other words, we can consider the changes in the flow to be brought about by waves that bring a specific change and travel in a specific direction.

Unfortunately the steady compressible flow equations are no longer fully hyperbolic, except for the two-dimensional supersonic case.[5] In the subsonic case the long-term observation implied by the steady state would allow acoustic waves to travel everywhere, making the problem partly elliptic. The steady, subsonic compressible Navier-Stokes equations have a mixed hyperbolic-elliptic character. If we use an unsteady or at least pseudo-unsteady formulation, we can always consider the steady state to be reached in the limit of a transient process. This in turn uses the unsteady equations, which are fully hyperbolic. This is the typical approach that is taken when discretising the compressible flow equation and selecting boundary conditions.

5.2.7.2 The incompressible flow equations

Incompressible flow, whether unsteady or steady, is a different case. In the incompressible limit the time derivative of the continuity equation disappears and the mass-conservation equation becomes a constraint on the divergence of the flow field (cf. Eq. 2.6). As shown in Sec. 3.6.6 we can use continuity and

[5] Strictly speaking the equations are only fully hyperbolic in two-dimensional flow.

the momentum equation to derive an equation for the pressure, the pressure correction equation (3.23). This equation is similar in structure to the heat equation of Sec. 3.3 which is an elliptic equation. As was demonstrated in Section 5.2.2, elliptic equations do not possess a time-like direction. They cannot be solved by our favourite method, marching in time. Note that the fact that the time derivative is zero implies that the divergence constraint has to be satisfied exactly at each iteration.

The incompressible Navier-Stokes equations, hence, are a coupled system of equations with an elliptic equation for the pressure and two or three hyperbolic equations for the conservation of momentum. If energy exchanges are relevant, e.g., due to wall heating, a conservation equation for energy needs to be added. Section 5.2.6 discussed discretisation for incompressible flows in a bit more detail. The elliptic part and the hyperbolic part require different discretisations and need to be coupled carefully to ensure good convergence properties. CFD packages in general offer a choice of coupling methods: one of the classic ones and a good general purpose choice is the SIMPLE method. Similarly, boundary conditions for compressible or incompressible flows differ slightly as we respond to the different characters of the equations.

5.3 Choice of boundary conditions

The dashed lines in Fig. 5.6 indicate where boundary information needs to be specified. The hyperbolic case needs an initial condition at $t = 0$ and boundary conditions at $x = 0$ and $x = 1$. Boundary specified at either end of the domain require some time to be felt at the interior. The parabolic case requires an initial condition at $t = 0$ and boundary conditions at $x = 0$ and $x = 1$ for the past and the present. The future (or downstream if you wish to see it that way) has no influence and no boundary conditions need to be specified. The elliptic case requires boundary conditions around the entire domain at $x = 0, 1$ and $y = 0, 1$ since all points depend and influence each other.

5.3.1 Boundary conditions for incompressible flow

The incompressible flow equations are of mixed elliptic-hyperbolic type, hence, the boundary conditions need to be imposed accordingly. The momentum equation can be viewed as a transport equation for the velocity, a hyperbolic equation which requires an initial solution amd imposition of the incoming velocity at inlet boundaries. The velocity must not be imposed at the exit.

The pressure-correction equation which is used in most incompressible discretisations (see Sec. 3.6.6) is of the form of the heat equation which is an ellliptic equation. It needs a boundary condition at each boundary. This condition can take the form of a fixed value, e.g., zero gauge pressure at the outlet, which is called a *Dirichlet*[6] condition. It can also take the form of imposing the gradient, called a *Neumann*[7] condition, e.g., prescribing zero gradient of pressure normal to a solid wall.

As shown in Sec. 5.2.2, elliptic problems do not require an initial solution; the final solution is uniquely determined by the boundary conditions. Indeed, the absolute value of the pressure does not appear in the momentum equation. Only the pressure gradient is present. In practice, however, we want to investigate a physical pressure field which we obtain by integrating this pressure gradient from a known value, e.g., the zero pressure value at the exit or the pressure in a particular point that we define to be fixed at a particular value. This means the pressure needs to be fixed somewhere. Another aspect to be aware of is that a suitable choice of initial pressure field may significantly accelerate the convergence of the equations. So in practice all CFD solvers will ask for an initial pressure field.

The manual of your CFD code will offer help on the multitude of conditions that have been implemented, as well as how and when to use it. Some typical, basic boundary conditions are listed below. Their naming and exact implementation may differ from code to code, so check the manual.

Velocity Inlet Velocity inlet boundary conditions are used to define the velocity at inlet boundaries.

Velocity can often be specified by components, or e.g., by magnitude and direction, which can either be fixed or taken normal to the boundary. The velocity does not have to be uniform. Typically the user can specify a velocity profile, e.g., parabolic for fully developed pipe flow.

If there are additional transport equations in your model e.g., for energy and/or turbulence, then inlet values for temperature and/or the turbulence values also need to be specified.

Pressure Inlet Pressure inlet boundary conditions define the total pressure at flow inlets. Given the static pressure from inside the domain, this allows to calculate and impose incoming velocity.

Such a condition might be useful if the flow is connected to a large reservoir or an ambient room where the fluid has this total pressure.

Pressure Outlet Pressure outlet boundary conditions define the static pressure at an outlet. This works well if the pressure is known and the flow is exiting.

It is best not to have an exit plane if the flow is partly re-entering, e.g., due to a recirculation zone that crosses the exit plane. Some solvers allow to specify the total pressure of the flow that is coming back in, but often this state cannot be determined accurately enough.

---

[6]Peter Gustav Lejeune Dirichlet, German mathematician, 1805-1859.

[7]John von Neumann, Hungarian and American pure and applied mathematician and physicist, 1903-1957.

Wall Wall boundary conditions impose the velocity at the wall. Typical for viscous flow would be a no-slip condition such as zero velocity at a stationary wall. If the wall is moving, no-slip would imply that the flow is moving with the speed of the wall.

Slip-wall conditions are typical for inviscid flow and force the flow to be tangent to the wall. The flow cannot pentetrate the wall. This condition may also be useful in viscous flow for walls where the viscous effects are not of interest, and not simulating those effects may result in computational savings.

Symmetry Symmetry boundary conditions impose symmetry of the flow field. Simulating half an aeroplane with a symmetry plane in the middle results in half the mesh size. Naturally, the flow-field needs to be symmetric in this case; no yaw angle is allowed.

The behaviour of the velocity at the symmetry plane is the same as for a slip-wall: the velocity needs to be tangent to the symmetry plane. If propertly implemented, however, the symmetry plane condition will also enforce that the gradients of velocity or pressure are zero in the symmetry plane in the direction perpendicular to the plane.

Periodicity Periodic boundary conditions are appropriate if geometry repeats periodically, e.g., this allows to simulate only a single turbine blade on a rotor, rather than the entire rotor with many blades. The flow conditions at the periodic boundary pairs are then matched with an appropriate rotation and/or translation.

5.3.2 Boundary conditions for hyperbolic equations

In the case of the scalar equation (Eq. 2.35) the choice of boundary condition was clear. There is only one variable; hence, only one piece of information is transported. The direction on which information travels, the *characteristic* direction in space and time, is the advection speed, $a = \Delta x/\Delta t$. At the inlet one piece of information, in our examples the value of the passive scalar u, needs to be specified; at the outlet u must not be specified.

Similar reasoning applies to the compressible unsteady Euler and Navier-Stokes equations, except that not only one but 4 (2-D) or 5 (3-D) pieces of information are required to compute the state of the fluid at a point, e.g., in 2-D four pieces of information to determine ρ, u, v and p. The characteristics that carry the information in order to compute the state can in general also point upstream such as an acoustic wave travelling upstream with the speed of sound in a subsonic flow. The key to imposing boundary conditions correctly is to determine for each boundary how many pieces of information are entering — and impose conditions for exactly that many, and the right combination such that those entering values can be determined — and to not impose any conditions for the remaining outgoing characteristics.

For a subsonic inlet this means that three quantities need to be imposed, such as velocity and density. A subsonic outlet imposes one quantity, e.g., pressure. In a supersonic inlet all 4 elements of the state need to be specified; none must be specified at the supersonic outlet.

Compressible flow discretisations typically allow a *far-field* condition, where the user specifies all the states of the incoming freestream flow. The flow solver then can evaluate how many waves are entering or exiting and impose the correct number of conditions.

5.4 Exercises

5.1 Consider the case of simulating turbulent flow over a car in a box-shaped test-section of a wind-tunnel. The inlet velocity is known as 30 m/s.

What boundary conditions are assigned at the test-section's in- and outlet, the test-section walls and the vehicle surface? Can the cost of the computation be reduced with an appropriate boundary condition?

5.2 An aerofoil is simulated in freestream, inviscid flow. The outer boundary of the grid is placed 5 chord lengths away from the aerofoil. Uniform freestream velocity is imposed at the outer boundary; tangency (slip wall) is imposed at the wall.

What are the problems with this setup for boundary geometry and conditions?

5.3 Flow over an ascending rocket for a satellite launch is simulated to determine the drag. At this stage in the flight, the rocket's speed is supersonic with Ma=2. The domain is cylindrical around the axis of the rocket.

What boundary conditions need to be imposed at the various surfaces?

5.4 You are solving the heat equation on a rectangular plate. The left end of the plate is heated to 400 K; the right end is cooled to 300 K. The top and bottom are insulated.

What boundary conditions are needed at which sides?

5.5 The SIMPLE discretisation for the incompressible Navier-Stokes equations (cf. Sec. 3.6.6) solves a Poisson equation for the pressure. (Recall that Poisson equations and Laplace equations have elliptic behaviour.)

Consider the simulation of Exercise 5.1. What boundary conditions are needed to be imposed for solution of the pressure correction equation?

6

Turbulence modelling

6.1 The challenges of turbulent flow for CFD

Reynolds'[1] well-known experiment, a description of which can be found in most general fluid dynamic textbooks, investigated flow in a circular pipe. Dye was injected at the centreline which allowed visualisation of the behaviour of the flow. He found laminar flow at low speeds with the dye forming a linear trail that very gradually diffused as it was transported downstream. At slightly higher speeds he could observe intermittent bursts of irregularity that washed out the dye trail through strong mixing with the surrounding fluid, but these disturbances disappeared shortly after their generation. This regime is called transitional flow: at times perturbations become amplified and generate a burst of chaotic, swirling motion, but these *turbulent spots* are not sustained and the perturbation dies out. At higher speeds again Reynolds observed the flow to become fully chaotic: the dye trail very quickly became mixed into the fluid by sustained chaotic motion. This is *turbulent flow.*

Turbulence is characterised by the existence of *eddies*, vortical swirls in the fluid, in a variety of *scales* (sizes). Fig. 6.1 shows a flow with a range of eddies in different sizes viewed from space. The satellite image shows Alexander Selkirk island in the southern Pacific ocean. The island's summit peaks 1600 m above sea level and air in the lower atmospheric levels has to flow around the obstacle. The upper level of cloud cover on the day the image was taken is well below the summit; hence, the clouds help visualise the pattern of vortical motion of the flow around the island.

The flow pattern that can be observed is called a *von Karman*[2] *vortex street.* The flow is not symmetric downstream of the island but pathlines separate from the island's contour further upstream on one side, while on the other side the flow remains attached longer and curves around behind. This leads to a swirling motion, an eddy, in the flow on the separated side. This eddy will increase in size until it becomes too large to remain attached to the island, detaches and then is transported downstream with the flow. The pattern then repeats with an eddy growing on the other side. These initial eddies are the largest size we can observe in Fig. 6.1.

[1]Osborne Reynolds, 1842–1912, British fluid dynamicist.

[2]Theodore von Kármán 1881–1963, a Hungarian-American aerospace engineer.

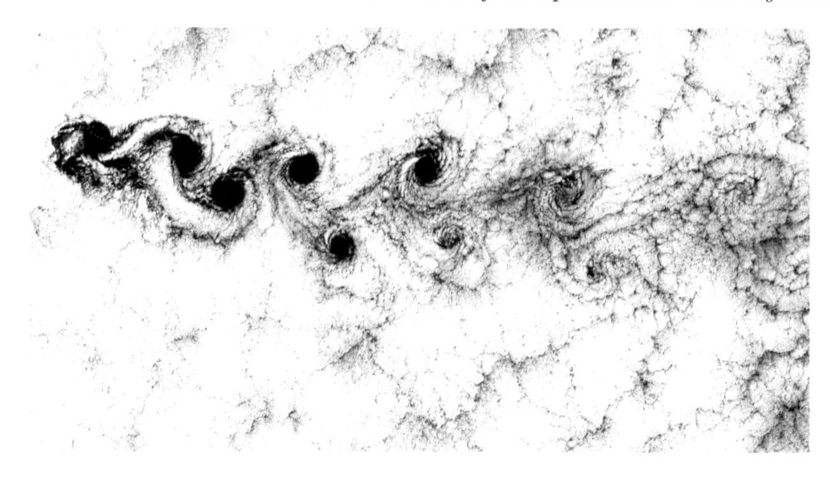

FIGURE 6.1
Cloud patterns around Alexander Selkirk island visualise the eddy motion of a von Karman vortex street downstream of the island. (Courtesy: NASA)

Of more interest to us is the range of scales we can observe in the wake of the island. The initial eddy that is generated by the island is strongly two-dimensional with its rotational axis normal to the water surface. It is marked by the black cloudless zones at the cores of each vortex. These largest vortices do not remain very ordered, however. We can see that as the vortex travels downstream its motion becomes more chaotic; smaller eddies form between the larger ones as seen by the ripples in cloud cover. The direction of those smaller vortices is not uniform but more chaotic. If we looked far enough downstream, not shown in this image, we would observe that all the larger eddies are broken down into smaller and smaller ones. Those smaller eddies would still have a range of sizes but their orientation would be fully anisotropic with no preferred direction.

This *cascade* of vortices is a major phenomenon in turbulent flow.

- Eddies can be created either by growth of perturbations due to the instability of laminar flow at high Reynolds numbers or e.g., due to unsteady separation as in the von Kármán vortex street behind a cylinder.

- These large eddies themselves are not stable in turbulent flow, but break down into smaller ones, which in turn break up, and so on.

- Eddies of different sizes interact in fully turbulent flow and turbulent kinetic energy is transferred between the different scales. The transfer of energy from the largest to the smallest scales is called the *Kolmogorov cascade.*

The size of the largest eddies is mainly determined by the geometry of the flow, e.g., the size of the island in Fig. 6.1, or the thickness of the layer in boundary layer flow. The kinetic energy in the rotation of the eddy is taken from the mean flow. As the eddies break up and form smaller ones, there is a flux of turbulent kinetic energy transferred from the large eddies to smaller and smaller ones. Smaller eddies can at rare instances combine to form bigger ones and hence transfer energy to larger scales, but the net flux of turbulent energy is toward the smaller scales.

Very importantly, there is a lower size limit for the eddies, i.e., a bottom end of the *Kolmogorov cascade.*[3] As their size decreases, so does the Reynolds number based on the eddy diameter which is the ratio of inertial forces over viscous forces acting on the eddy. At larger Reynolds numbers inertial forces dominate and viscous effects are nearly negligible: all the kinetic energy is passed from the larger to the smaller scales. Viscous losses can be neglected. This range of scales is hence called the *inertial range.* The inertial forces diminish as the eddy size decreases. The viscous forces on the other hand remain of constant order. As the Reynolds number (based on eddy diameter) reaches unity from above, the viscous forces will become dominant and the vortex will be dissipated; turbulent action will cease at that scale.

6.2 Description of turbulent flow

The equations for laminar flow can be solved analytically in a variety of cases, but we know only local analytic approximations to turbulent flow.

Turbulence is unsteady and three-dimensional by nature; hence, the steady or two-dimensional equations can't describe turbulence correctly. The delicate balance of turbulent kinetic energy in that cascade, however, is described at all scales by the unsteady 3-D Navier-Stokes equations: turbulent flow still has to satisfy continuity and balance of momentum. You will have seen in your earlier studies of fluid dynamics that analytic solutions for the equations only exist for very few simplified cases, so it will come as no surprise that we cannot (yet) solve the turbulent flow equations.

On the other hand Sec. 4.2 has shown that if we have a consistent discretisation and make the mesh fine enough, discretisation errors will become negligible. Hence, to accurately simulate turbulent flow we could try to exploit the fact that there is a smallest turbulent scale and make the mesh and time-step fine enough to accurately resolve these smallest eddies in space and time. Unfortunately, for relevant Reynolds numbers in industrial application, the smallest eddies are so small that we won't have the computers to run the required mesh sizes and numbers of iterations for a very long time. This approach, termed *Direct Numerical Simulation,* is feasible for smaller Reynolds numbers and academic problems investigating fundamentals of turbulence.

[3]Andrey Nikolaevich Kolmogorov, 1903–1987, Soviet mathematician.

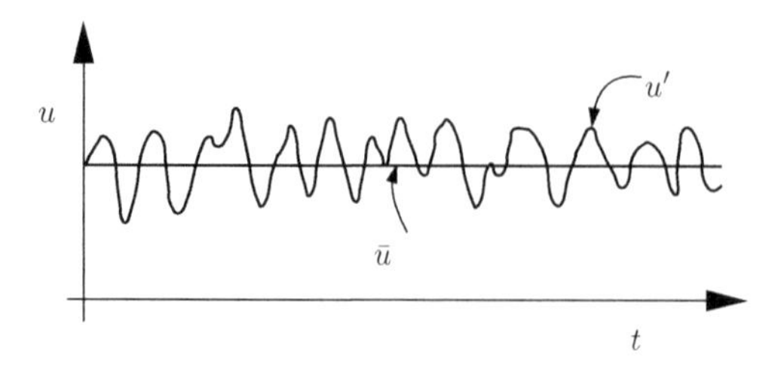

FIGURE 6.2
Turbulent velocity fluctuation u' around a steady average $\bar{u}$.

But this approach is not feasible for industrial application and we shall not discuss this here further.

The fluctuations of turbulence at the smaller scale are of no practical interest in most engineering problems. When the engineer e.g., computes the drag over a car, he/she cares for quantities averaged over long periods of time compared to the frequency of the fluctuations. When measured with a probe that can resolve high frequencies, such as e.g., a hot-wire anemometer, one finds that turbulent flow exhibits high frequency fluctuations around a mean velocity. If the average flow is steady, the mean velocity $\bar{u}$ is constant as shown in Fig. 6.2.

The unsteady, or *instantaneous* velocity u can be decomposed into a steady part, the *average velocity* $\bar{u}$, and a *fluctuation* u':

$$u = \bar{u} + u', \tag{6.1}$$

with $\bar{u}$ being the time-average of u over a sufficiently long time T:

$$\bar{u} = \frac{1}{T}\int_{t=0}^{t=T} u dt. \tag{6.2}$$

One can hence consider the averaged effect of turbulence over time, which would be similar to measuring the static and dynamic pressures in the flow with slow-acting U-tube manometers. In this way we can extract 'steady' measurements from turbulent flow, or consider a 'steady turbulent flow field' $\bar{u}$ which may be two-dimensional.

Similarly, the shear stress can be decomposed into a part due to the laminar shearing τ_l and due to the turbulent motion τ_t, $\tau = \tau_l + \tau_t$. The laminar shear stress in Newtonian fluids is proportional to the rate of strain, $d\bar{u}/dy$; the proportionality constant is the viscosity μ which is a fluid property. .

The turbulent shear stress τ_t is a property of the flow, not the fluid. To estimate τ_t consider the interface of area dA between two fluid elements in inviscid flow sketched in Fig. 6.3. The upper element is moving with velocity $u + u', v'$ in a uniform field with mean velocity u, v with $v = 0$. Consider the case where $u' > 0$ and $v' < 0$: the upper fluid element moves faster in x and

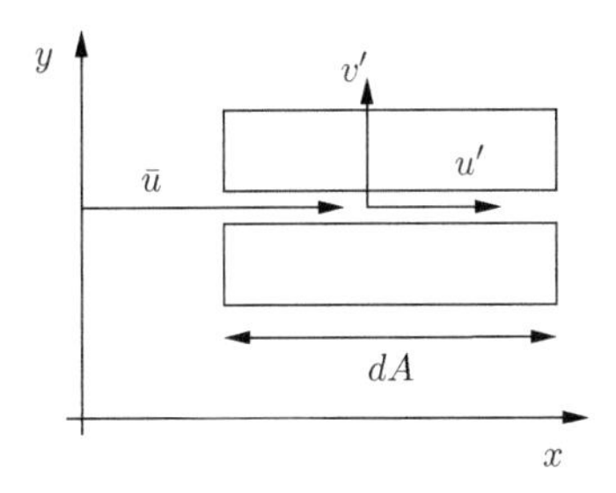

FIGURE 6.3
Momentum transport by turbulence.

downward in y. In the lower element the fluid has an x-momentum of ρu, in the upper one of $\rho(u+u')$. The downward movement of the upper fluid will bring faster fluid into the lower element, increasing the average x-momentum in the lower element. The volume of fluid in unit time across the interface between the two fluid elements is $v'dA$. The x-momentum flux across the interface is consequently

$$f_x = \rho u v' dA = \rho(\bar{u} + u')v' dA = \rho \bar{u} v' dA + \rho u' v' dA. \tag{6.3}$$

The momentum flux averaged over time then is

$$\bar{f_x} = \rho \bar{u} \overline{v'} dA + \rho \overline{u'v'} dA = \rho \overline{u'v'} dA. \tag{6.4}$$

The time average of the fluctuations vanish, $\overline{u'} = \overline{v'} = 0$, by the definition of the average in Eq. 6.2 and the fluctuation in Eq. 6.1. The average of the correlation $\overline{u'v'}$ does not vanish.

As expected the turbulent momentum exchange produces an accelerating force on the lower element if $u' > 0, v' < 0$. Following the sign conventions for shear forces, this corresponds to

$$\tau_t = -\rho \overline{u'v'}, \tag{6.5}$$

the *Reynolds stress.* The Reynolds stress terms describe a dissipative effect; velocity differences are evened out. However, it is not a viscous effect as can be seen from the absence of a viscosity coefficient in (6.5).

This equation for the Reynolds stress can also be derived by inserting the velocity description $u = \bar{u} + u'$ into the differential equations and averaging as shown in Section 6.5.2. The correlation $\overline{u'v'}$ is the statistical average value of the product of the two fluctuation velocities. This value is highly dependent on many flow quantities, and is in general not only dependent on averaged flow state $\bar{u}$. However, reasonable models for it have been proposed that adequately describe its behaviour, the most popular ones being actually based on the averaged state only.

6.3 Self-similar profiles through scaling

6.3.1 Laminar velocity profiles

Fully developed laminar flow in pipes, Hagen[4]-Poiseuille[5] flow, is one of the flows where we have an analytic solution to the Navier-Stokes equations. The velocity profile is parabolic, with zero velocity at the pipe wall and a peak velocity u_{max} at the centreline. All fully developed laminar velocity profiles in pipes can therefore be described with the velocity profile of

$$\frac{u(r)}{u_{max}} = \left(1 - (\frac{r}{R})^2\right)$$

with the pipe radius R and r the radial position in the pipe. By scaling the velocity and the radius appropriately, we can obtain a universal velocity profile that is valid for all diameters and speeds, as long as the flow is laminar and fully developed. Such a profile is called *self-similar* as with appropriate scaling any profile will match the universal one.

Similarly we can derive a self-similar velocity profile for a fully developed laminar boundary layer with zero pressure gradient by scaling the distance from the wall y with the boundary layer thickness δ and the velocity u with the freestream velocity u_∞ outside the boundary layer,

$$u_{sc} = \frac{u}{u_\infty} = f\left(\frac{y}{\delta}\right) = f(y_{sc}).$$

all velocity profiles for fully developed zero pressure-gradient boundary layers scaled in this way will fall on top of each other; the scaled velocity u_{sc} is only a function of the scaled wall distance y_{sc}.

6.3.2 Turbulent velocity profile

Self-similar velocity profiles also exist for turbulent flows. Typical velocity profiles in turbulent pipe flow are sketched in Fig. 6.4. It can be observed that for increasing Re the profile becomes steeper near the pipe wall and flatter near the centre. The velocity profiles can be approximated by

$$\frac{u(r)}{u_{max}} = \left(1 - \frac{r}{R}\right)^n. \tag{6.6}$$

Prandtl[6] used the semi-empirical formula by Blasius for $2000 < Re < 10^5$ to derive an exponent of $n = 1/7$. In general, though, n is a function of Re: $n = n(Re)$, e.g., $n = 1/6$ is an acceptable approximation to measurements for

[4]Gotthilf Heinrich Ludwig Hagen, 1797-1884, German physicist and hydraulic engineer.
[5]Jean Léonard Marie Poiseuille, 1797–1869, French physicist and physiologist.
[6]Ludwig Prandtl, German fluid dynamicist, 1875–1953.

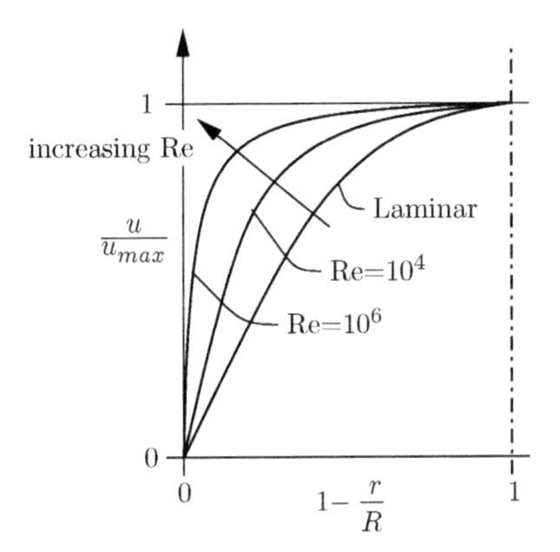

FIGURE 6.4
Velocity distributions of turbulent flow in smooth pipes.

$Re = 4 \times 10^3$ and $n = 1/10$ for $Re = 3.2 \times 10^6$. As is evident, this simple scaling, which is extended from ideas of laminar flow, is not universal, but does depend on the Reynolds number. This aspects makes it not very useful in general.

It can also be observed that this velocity distribution is not valid for the regions very close to the wall. For any exponent $n < 1$, which includes the useful values of $\frac{1}{4} \le n \le \frac{1}{10}$, the velocity gradient at the wall becomes infinite: $(du/dr)_{r=R} = -\infty$. That is, even if an appropriate value of n can be found, this type of profile cannot be used to approximate the velocity gradient at the wall.

This is a fatal shortcoming in CFD where we need to approximate the wall shear stress in order to compute the momentum balance. The wall shear stress depends directly on the velocity gradient at the wall, which is not available through this scaling. In the next section we will consider an alternative scaling that focuses on the near-wall behaviour of the flow.

6.4 Velocity profiles of turbulent boundary layers

Fig. 6.5 recalls the key concepts of boundary layers; for more details see the typical textbooks. To simplify the concepts, we consider as an example flow over a flat plate which provides a fixed stagnation point at which the boundary layer starts. The boundary layer is initially laminar. Transition to turbulence occurs around a critical Reynolds number Re_{crit} of around 50000, based on length from the leading edge. It is very difficult to model transition of boundary layers so in practice in CFD boundary layers are simulated as turbulent

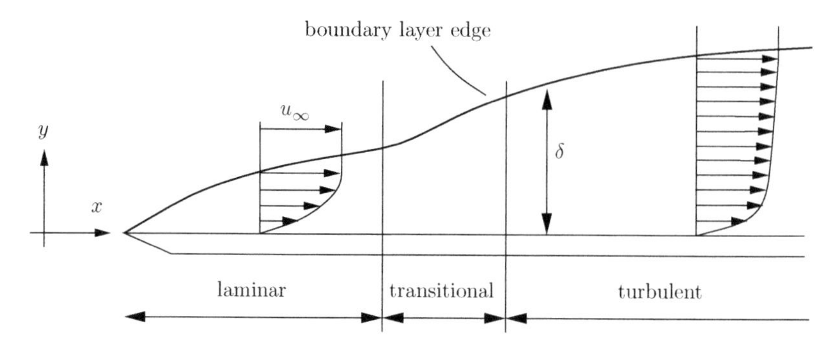

FIGURE 6.5
Boundary layer on a flat plate.

from the beginning. This approximation is generally good enough for aeronautical applications but clearly finds its limits when using CFD to simulate modern drag reduction techniques using laminar flow.

Key parameters of interest in CFD are the boundary layer thickness $\delta(x)$, the velocity profile $u(y)$, and, most importantly, the wall shear stress $(\tau_w(x)$. The boundary layer thickness is defined as the height above the plate where the profile reaches 99% of the freestream velocity outside the boundary layer. As the sketch shows, the boundary layer grows along the plate.

In the laminar region in Newtonian flow with constant viscosity μ, the wall shear stress is directly proportional to the velocity gradient normal to the wall,

$$\tau_0 = \mu \left. \frac{\partial u}{\partial y} \right|_0 ,$$

where y is the direction normal to the flat plate and the subscript 0 indicates the wall with $y = 0$. As we have seen in Sec. 6.3.1, there is a self-similar profile for fully developed laminar boundary layers which scales the velocity u with the freestream velocity u_∞, which remains constant, and the height y with the boundary layer thickness $\delta(x)$, which varies with x. As a consequence, the velocity gradient and wall shear stress at the wall decrease with x. At the stagnation point this scaling would produce zero thickness and infinite wall shear stress, but since the boundary layer is not fully developed in the stagnation point, the scaling is not valid there.

6.4.1 Outer scaling: friction velocity

Obtaining wall shear stress for the fully turbulent region is more difficult. It is typically approached by considering the x- and y-momentum equations of the Navier-Stokes equation and eliminating terms that are much smaller in order of magnitude in the flat plate boundary layer. Remaining terms then often are estimated using dimensional analysis.

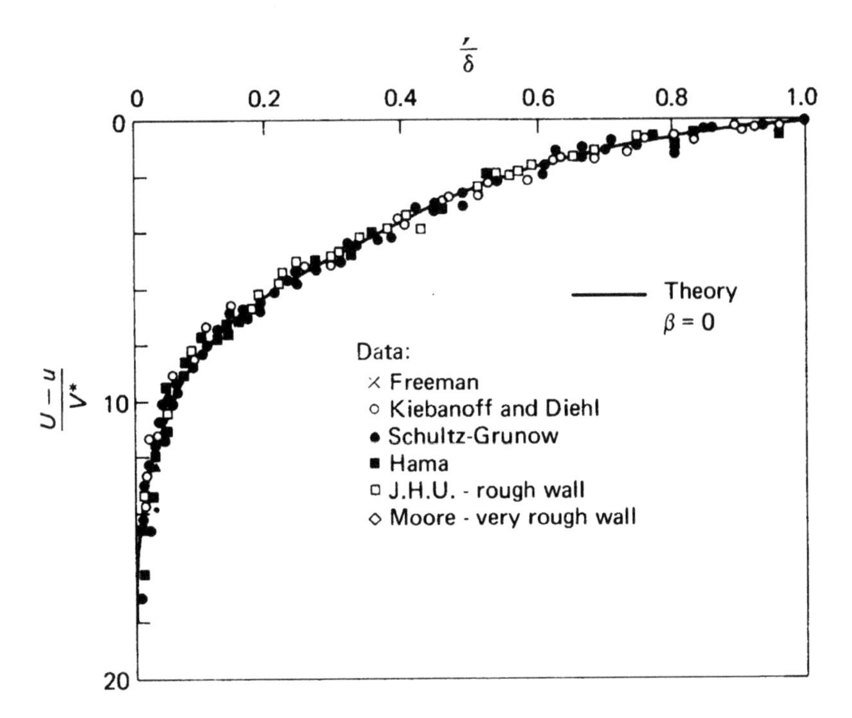

FIGURE 6.6
Velocity profile in a turbulent boundary layer from [6].

The layer from around 20% of the boundary layer thickness up to the edge of the boundary layer is called the *outer layer*. The loss of velocity in the boundary layer depends on all flow parameters,

$$u_\infty - u = f\left(\tau_0, \rho, y, \delta, \frac{\partial p}{\partial x}\right). \tag{6.7}$$

The analysis suggests for this outer region of the boundary layer an *outer scaling* which scales wall distance y as before with boundary layer thickness δ, but different from Eq. 6.6 it scales the velocity loss $u_\infty - u$ with the friction velocity u^* or u_τ (both symbols are used in the literature) as $u_\tau = \sqrt{\frac{\tau_0}{\rho}}$. The derivation of the friction velocity will demonstrated with the presentation of the inner scaling in the next section (6.4.2). Fig. 6.6 shows how the measurements map onto this self-similar profile when this scaling is used.

Rather complex but quite accurate formulae can be derived to express this profile in the form of $u = f(y)$. However, this profile is accurate only for the outer regions of the flow. It will give poor predictions of the shear stress at the wall. Most importantly, the wall shear stress τ needs to be known to compute the friction velocity u_τ in the first place. Hence, we need to develop this further to describe the near-wall behaviour.

6.4.2 Inner scaling: non-dimensional wall distance y^+

6.4.2.1 Linear sublayer

As previously described, turbulent flow exhibits a cascade of eddies from largest sizes down to sizes at which viscosity becomes dominant over inertia. In the very close proximity to the wall, of the order of the size of the smallest scales, there cannot be any turbulent eddies as they are being damped by wall friction; the flow becomes locally laminar. In Newtonian flow the shear stress then becomes

$$\tau \approx \tau_0 = \mu \frac{\partial u}{\partial y}. \tag{6.8}$$

Integrating this equation with respect to the wall distance y one finds

$$\int \tau dy = \tau_0 y = \mu u = \rho \nu u. \tag{6.9}$$

The force balance over a fluid element in this layer close to the wall has to be zero since we consider fully developed flow: pressure is constant as we assumed $\frac{\partial p}{\partial x} = 0$. As a consequence, the shear stress also has to be constant as found in Couette flow: close to the wall the velocity profile has to be linear.

Similarly as found for the outer region, we seek a non-dimensional approximation that is universally valid very close to the wall in all boundary layers, subject to appropriate scaling. To non-dimensionalise the velocity, dimensional analysis offers the *friction velocity* u_τ,

$$u_\tau = \sqrt{\frac{\tau_0}{\rho}}. \tag{6.10}$$

Equation (6.9) then becomes

$$\frac{u}{u_\tau} = \frac{u}{\sqrt{\frac{\tau_0}{\rho}}} = \frac{y\sqrt{\frac{\tau_0}{\rho}}}{\nu} = \frac{y u_\tau}{\nu}. \tag{6.11}$$

Introducing the new variables u^+ and y^+ one finds the remarkable equation:

$$u^+ = \frac{u}{u_\tau} = \frac{y u_\tau}{\nu} = y^+. \tag{6.12}$$

That is, in inner scaling the velocity and the wall distance are directly proportional to each other: $u^+ = y^+$. This relation holds very close to the wall, $0 \leq y^+ \leq 5$. The relation between velocity u^+ and wall distance y^+ is linear; hence, this region is called the *linear sublayer*. It is also called the *viscous sublayer* as the shear stress in this region is produced entirely by molecular viscosity and not by turbulent mixing.

6.4.2.2 Log-layer

One can identify a zone further away from the wall but still well inside the inner regions of the boundary layer, say $30 \leq y^+ \leq 500$, where turbulence is present but the freestream velocity is not felt by the eddies there as they are buried down in the boundary layer too far from the outer edge. The velocity profile is dominated by the exchange of momentum through turbulent mixing (cf. Fig. 6.3), not by the influence of molecular viscosity (cf. Sec. 6.2). The velocity in the boundary layer then is a function of of all the local flow parameters that are also relevant to the outer scaling in Eg. 6.7, but the location is deep enough inside the boundary layer that the local scales are not influenced by the velocity outside the boundary layer,

$$u = f(\tau_0, y, \mu, \rho) \neq f(u_\infty). \tag{6.13}$$

That is, the velocity depends on the wall shear stress, the wall distance, the viscosity and the density, but not the freestream velocity.

Using again u_τ to non-dimensionalise the velocity profile, there is only one possible combination of the terms on the right-hand side to produce a non-dimensional group:

$$u^+ = \frac{u}{u_\tau} = f\left(\frac{y\sqrt{\frac{\tau_0}{\rho}}}{\nu}\right) = f(y^+). \tag{6.14}$$

Applying integration reveals that f must be a logarithmic function and is hence called the *log-law*. The layer where it is valid between $30 \leq y^+ \leq\approx 500$ is called the *log-layer*.

$$u^+ = \frac{1}{\kappa} \ln y^+ + B. \tag{6.15}$$

This layer is also called the *inertial sublayer* as dimensional analysis and experiments show that the shear stress in this layer is almost entirely due to turbulent mixing; effects of molecular viscosity are negligible. In turn this means that the kinetic energy of the eddies is not dissipated into heat, but mixed with the flow and hence preserved (hence the name of 'inertial sublayer'). But in the process of transfer, energy is passed from larger to smaller and smaller eddies until it does reach the smallest scales (and in boundary layers the near-wall regions) where viscous effects do take place.

Experiments determine $\kappa = 0.4$ and $B = 5.5$ for smooth walls. Together the linear sublayer and the log-law layer form the inner layer. The two curves and experimental measurements are shown in Fig. 6.7 taken from [6].

The experimental data plotted clearly confirm that in the inner scaling the boundary layer profile is independent of the Reynolds number.

In the linear sublayer we found in Sec. 6.4.2.1 that the wall shear stress τ is constant. We can extend this analysis to the include the log-layer. A fluid element in this region will not be subject to pressure differentials. The shear stress acting on the element may be resulting from molecular viscosity (we

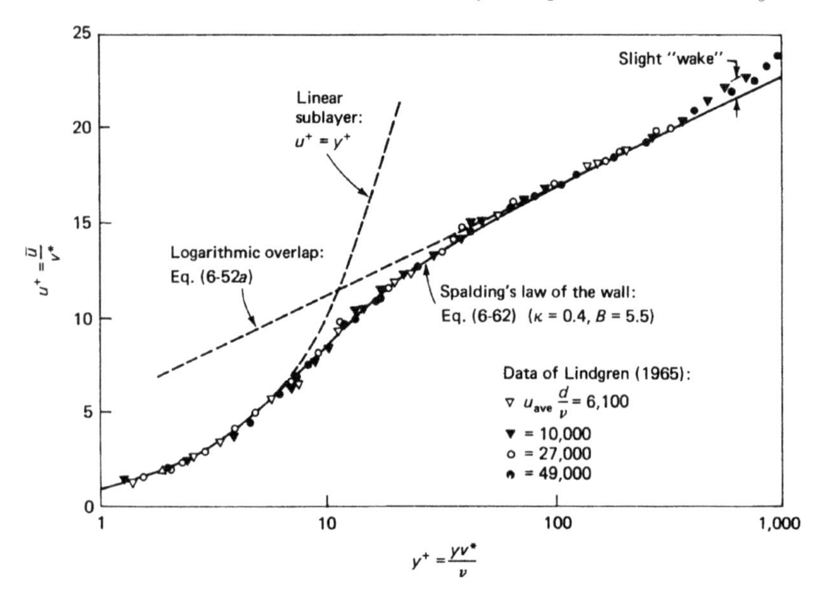

FIGURE 6.7
Linear sublayer and log-law confirmed by experiments of turbulent boundary layers from [6].

will see shortly that this term is actually negligibly small in the log-layer) and from shear stress due to turbulent mixing. If we take the assumption proposed by Boussinesq[7], that turbulent stresses are also proportional to the velocity gradients, which is a reasonable assumption in the lower layers of the boundary layer, we obtain

$$\tau_T = \mu_T \frac{\partial u}{\partial y}$$

where the subscript T refers to turbulence quantities. The *turbulent viscosity* or *eddy viscosity* μ_T is a property of the flow. It will not be a constant.

The shear forces acting on the fluid element are then

$$\tau = \tau_\mu + \tau_T = (\mu + \mu_T) \frac{\partial u}{\partial y}$$

Similar to how we proceeded for the linear sublayer, we can derive the momentum balance over the element which also for this case shows that the sum of all components of the shear stress will be constant in the y-direction in all of the layers very close to the wall, including the log-layer.

$$\frac{\partial}{\partial y}(\tau_\mu + \tau_T) = \frac{\partial}{\partial y}\left((\mu + \mu_T)\frac{\partial u}{\partial y}\right) = 0.$$

[7] Joseph Valentin Boussinesq, 1842–1929, French mathematician and physicist.

In this general form we see that both the molecular viscosity μ as well as the *turbulent viscosity* μ_T play a role and that throughout the inner layers the sum of the shear stress is constant in zero pressure-gradient boundary layers.

This actually is only the case in the *buffer layer*, the layer from $5 \leq y^+ \leq 30$ which is between the linear and the log layers. At its lower end $y^+ = 5$ joins the linear or viscous sublayer where turbulence is dissipated due to wall proximity and all the shear stress is due to molecular viscosity. At the upper end of the buffer layer at $y^+ = 30$ it joins the log-layer or inertial sublayer where viscous effects become negligible.

However, the fact that the overall shear stress is constant from the linear sublayer through to the log-layer is a most useful property and CFD exploits this to reduce the cost of resolving the near-wall region using wall-functions, as discussed in Sec. 6.7.0.3

6.5 Levels of turbulence modelling

As described in Sec. 6.2, turbulent flows exhibit a cascade of vortices, eddies, from largest scales defined by the geometry of the flow to the smallest sizes which depend on the Reynolds number. Turbulent flows are inherently unsteady, three-dimensional and contain very small scales at high Reynolds numbers. Resolving these small length and timescales in three dimensions with a fine mesh is not realistic for the foreseeable future for industrially relevant flows. In the following we will discuss this 'brute force' approach, as well as more practicable approaches that model the turbulence rather than resolve it.

6.5.1 Direct Numerical Simulation (DNS)

The most stringent approach to simulation of turbulence is to adequately resolve all the turbulent scales, from the largest scales to the smallest Kolmogorov scales where dissipation takes hold. This requires to choose the smallest mesh scale such that the smallest eddy is represented by an appropriate number of mesh points, as well as choosing the time-step small enough that the rotation of the eddy is discretised with an acceptably small truncation error. This method is called *Direct Numerical Simulation, DNS.*

With today's computer hardware and numerical algorithms, it is only possible to simulate the entire turbulent cascade for very small geometries and limited Reynolds numbers. Many studies have simulated, for example, homogeneous turbulence in a square box. One of the two main problems is that in general artificial viscosity is required for the stability of the numerical method (cf. Sec. 4.3.1). The combined viscosity of the mathematical model and its numerical implementation is too large and does not behave as in the mathematical model. The critical balance of energy exchange along the cascade

	Direct Numerical Simulation (DNS)
++	no modelling of turbulence, resolves the full turbulent spectrum.
−−	requires extremely fine mesh and timescales, only affordable for small Reynolds numbers even when using huge computational resources.
→	often used for basic simulation of physical phenomena or for validating LES.

TABLE 6.1
Summary of pros and cons of turbulence simulation with DNS.

is not accurately represented since there is too much viscosity affecting the smallest scales. DNS hence are implemented on specialised grids using very high-order numerical methods with low artificial viscosity. These numerical methods are not applicable to the complex industrial geometries.

But more importantly, at even moderate Reynolds numbers the length scales at which a local Re of unity is achieved become very small. To accurately simulate the nonlinear effects of breakup of eddies and their viscous demise, one has to represent even the smallest eddies with a number of mesh points. The mesh sizes required for DNS at moderate Reynolds numbers on realistic geometries at the moment exceed the capacities of even the most powerful computers. If Moore's law is to hold and the performance and capacity of computers continues to grow at the current rate, predictions show that we have to wait well into the next millennium for a DNS over a passenger jet at cruise conditions.

6.5.2 Reynolds-Averaged Navier-Stokes (RANS)

The most common way to solve for turbulent flows is to model the entire range of turbulent fluctuations. Consider the conservation equation for the momentum in x for two-dimensional incompressible flow:

$$\frac{\partial u}{\partial t} + u\frac{\partial u}{\partial x} + v\frac{\partial u}{\partial y} = -\frac{1}{\rho}\frac{\partial p}{\partial x} + \frac{\partial}{\partial x}\left(\nu\frac{\partial u}{\partial x}\right) + \frac{\partial}{\partial y}\left(\nu\frac{\partial u}{\partial y}\right). \tag{6.16}$$

The rate of change of the u-velocity, $\frac{\partial u}{\partial t}$, depends on the convective transport of momentum, $u\frac{\partial u}{\partial x}$ and $v\frac{\partial u}{\partial y}$, the pressure gradient, $\frac{1}{\rho}\frac{\partial p}{\partial x}$ and the viscous dissipation, $\frac{\partial}{\partial x}\left(\nu\frac{\partial u}{\partial x}\right)$ and $\frac{\partial}{\partial y}\left(\nu\frac{\partial u}{\partial y}\right)$. In an engineering context one is most interested in long-time averages of the flow quantities.

The velocity is decomposed into an average and a fluctuating component, $u = \bar{u} + u'$, shown in Fig. 6.2. Having defined $\bar{u}$ as the long-time average, the average of the fluctuations $\overline{u'}$ is 0:

$$\overline{u} = \frac{1}{T}\int_t^{t+T} u dt \tag{6.17}$$
$$= \frac{1}{T}\int_t^{t+T} \overline{u} + u' dt \tag{6.18}$$
$$= \frac{1}{T}\int_t^{t+T} \overline{u} dt + \frac{1}{T}\int_t^{t+T} u' dt \tag{6.19}$$
$$= \overline{u}\frac{1}{T}\int_t^{t+T} dt + \frac{1}{T}\int_t^{t+T} u' dt \tag{6.20}$$
$$= \overline{u}. \tag{6.21}$$

The x-momentum equation reads with this definition of u

$$\begin{aligned} &\frac{\partial(\overline{u}+u')}{\partial t} + (\overline{u}+u')\frac{\partial(\overline{u}+u')}{\partial x} + v\frac{\partial(\overline{u}+u')}{\partial y} = \\ &\quad \frac{1}{\rho}\frac{\partial(\overline{p}+p')}{\partial x} + \frac{\partial}{\partial x}\left(\nu\frac{\partial(\overline{u}+u')}{\partial x}\right) + \frac{\partial}{\partial y}\left(\nu\frac{\partial(\overline{u}+u')}{\partial y}\right). \end{aligned} \tag{6.22}$$

Taking the long-time average of these equations and simplifying with the identity $\overline{u'} = \overline{v'} = \overline{w'} = 0$, one finds the *Reynolds-averaged* equations:

$$\begin{aligned} \frac{\partial\bar{u}}{\partial t} + \bar{u}\frac{\partial\bar{u}}{\partial x} + \bar{v}\frac{\partial\bar{u}}{\partial y} &= \frac{1}{\rho}\frac{\partial\bar{p}}{\partial x} + \frac{\partial}{\partial x}\left(\nu\frac{\partial\bar{u}}{\partial x}\right) + \frac{\partial}{\partial y}\left(\nu\frac{\partial\bar{u}}{\partial y}\right) \\ &\quad + \left[-\frac{\partial\overline{u'u'}}{\partial x} - \frac{\partial\overline{u'v'}}{\partial y} - \frac{\partial\overline{u'w'}}{\partial z}\right]. \end{aligned} \tag{6.23}$$

As one can see, one obtains a momentum equation very similar to the fully unsteady equation (6.16), but augmented by terms that contain *first-order correlation* or *first-order moments* such as $\overline{u'v'}$. These correlations are averages of products of fluctuations and arise from the non-linear product $u\frac{\partial u}{\partial x}$; they are in general non-zero. The term $u'u'$, the square of the u-velocity fluctuations, is always positive; hence, its average is non-zero unless u' is always zero. The term $\overline{u'v'}$ is often negative, due to a conservation argument: mass conservation has to hold at any moment, so if u is increased locally, one often finds a decrease in the other velocity components v, w. These terms can be conveniently grouped with the velocity gradients and have the dimension of a (shear-)stress: they are called the *Reynolds stresses.* As shown in Section 6.2 they are not viscous terms since they don't carry a viscosity constant, but they indeed describe dissipative phenomena due to the exchange of momentum between layers.

One can derive conservation equations for these first-order correlations, but higher-order correlations such as $\overline{u'v'w'}$ then arise. Hence, at some level one has to start to model the fluctuations in order to close the system of equations. Most commonly the quantities like $\overline{u'v'}$ are modelled, and this

Reynolds-Averaged Navier-Stokes (RANS)	
++	all turbulent scales are modelled, which makes the model computationally very cheap
−−	all turbulent scales are modelled, which can incur significant errors for complex flows
+	simple flows predicted accurately, attached boundary layers and aligned free shear layers
+	a wealth of models to choose from with their own strengths and weaknesses which are well documented
→	the most commonly used approach, works acceptably well in many cases

TABLE 6.2
Summary of pros and cons of RANS turbulence modelling.

simple turbulence modelling works surprisingly well for flows with contained shear layers such as attached boundary layers. They fail dramatically for flows with large separation regions or unsteady flows.

Second-order closure methods, also called Reynolds stress equation models (RSM), actually do derive 7 additional transport equations for the 6 first-order correlations and for the dissipation rate ε. Second-order correlations are modelled. RSM gives better results in these cases, albeit at a substantially higher computational cost and a loss of stability of the numerical method.

6.5.3 Large Eddy (LES) and Detached Eddy Simulation (DES)

A possible reduction in computational effort is to fix an affordable finest mesh width and the corresponding time-step which resolves (i.e. numerically simulates the conservation equations) the part of the spectrum with the larger eddies down to a cut-off scale. The energy contained below the cut-off, and cascade and decay of the eddies smaller than this cut-off are then modelled (i.e. approximated) using a *sub-grid scale* model. These simulations are called *Large Eddy Simulations, LES.* While the physics of the larger, resolved eddies is captured accurately, the effect of the eddies of sub-grid scale are determined by the choice of sub-grid scale model. Note that the largest eddies carry most of the energy: practical mesh resolutions allow to simulate over 90 % of the turbulent kinetic energy. However, the final – and expensive – 10% still seem to be important in many cases, e.g. in boundary layer flows.

LES is no longer a research technology, commercial code vendors and industrial companies are making limited use of it in development, especially

Large Eddy Simulation (LES)	
+	resolves the physics of largest eddies containg a major fraction of the turbulent kinetic energy
–	models the smaller scales, which incurs modelling errors
+	allows a trade-off between computational cost and accuracy by adjusting the cut-off scale between resolution and modelling
+	much cheaper than RANS, using high-performance computing LES simulations of high-Re realistic industrial or atmospheric flows can be performed, detailed simulations at moderate Re are possible on typical computing hardware
+	often gives good or at least qualitatively correct answers where RANS fails, e.g. flow separation behind bluff bodies, combustion or weather forecasting
– –	to capture wall-bounded shear flows with current LES models requires a mesh resolution close to DNS. Accurate LES simulations of flows with strong shear layer effects are typically not affordable for industrial geometries at high Re.
→	very useful for moderate Re numbers or flows with limited wall effects

TABLE 6.3
Summary of pros and cons of turbulence modelling with LES.

when turbulent mixing needs to be predicted with accuracy as needed e.g. in gas turbine combustion or noise prediction.

A classic application for LES is weather forecasting where the Reynolds number can easily exceed 10^9. The mesh width of the models currently run for the country's daily forecast is of around a mile. All the phenomena below this scale are modelled. LES also has been very successfully applied to flows where the turbulent behaviour far from walls is most relevant, e.g. combustion in aero-engines or aircraft noise simulation.

LES is still a computationally very expensive simulation, run-times are a factor 10-100 times what is required by the methods that average all of the turbulent fluctuations. In particular the accurate resolution of wall-bounded flows near separation will require to resolve turbulent scales to a level close to the Kolmogorov scale , that is close to the mesh requirements of a DNS.

As an alternative the Detached Eddy Simulation (DES) method proposes to use LES in the field away from walls, and use a fully averaged RANS modelling approach (cf. Sec. 6.5.2) near the wall. The sub-grid scale viscosity close to the wall is then not estimated with a local model as done in LES,

Detached eddy simulation (DES)	
+	uses LES in the field away from solid walls to capture free turbulent eddy motion, and uses RANS near the wall to limit the required mesh resolution
+	easily implemented as an extension of the Spalart-Allmaras RANS model
+	if properly tuned, very good results can be obtained at moderate cost
−−	current DES formulations are very sensitive to the choice of switching length scale and mesh width. Fine tuning is required based on user knowledge which makes the approach not very robust for predictions for flows that the user is not familiar with
→	this is an active research area, further development is expected to make the DES approach more robust.

TABLE 6.4
Summary of pros and cons of turbulence modelling with DES.

but derived from a transport equation as in RANS approaches. While DES is very elegant in principle, the difficulty resides in the appropriate switching between RANS and LES. Current models appear still very sensitive to the choice of mesh width in the area where DES switches. Further progress on that is expected over the near future as this is a very active research area. With proper calibration of mesh width and cut-off scale, very good results at moderate computational cost can be achieved, but this does require very careful user intervention.

6.5.4 Summary of approaches to turbulence modelling

The tables 6.5.1-6.5.2 provide an overview over the different approaches. This book focuses on RANS models which is the most popular approach. However, LES and DES-based models which resolve some of the unsteady and three-dimensional turbulence are becoming feasible for some industrial applications, e.g., aeronautical and automotive, if high-performance computing is used.

This guide is mainly concerned with first-order closure which leads to models defining an *eddy viscosity* which augments the laminar viscosity, as discussed in the next section.

6.6 Eddy viscosity models

In both the RANS and the LES approaches all or part of the turbulent motion is averaged: these fluctuations are not present in our solution but of course lead to additional mixing, hence additional viscosity. Since we don't resolve these fluctuations, we need to model their effects using simplified physics.

Boussinesq, a French mathematician, suggested already in 1877 to model the turbulent shear stress as a function of the mean velocity gradient very similar to the laminar shear stress:

$$\tau_t = \mu_T \frac{d\bar{u}}{dy}. \tag{6.24}$$

The total shear stress then becomes

$$\tau = \tau_l + \tau_t = (\mu + \mu_T)\frac{d\bar{u}}{dy}. \tag{6.25}$$

Note the differences in μ. While μ, the molecular viscosity, is a fluid property which mostly depends on temperature, μ_T is a property of the flow. It is highly non-linear and can reach values in excess of 1000 times the value of μ in highly turbulent flows.

The Boussinesq approximation is not a wholly unreasonable thing to do. Experimental evidence shows that a mean flow gradient is required to sustain turbulence. In uniform flows, on the other hand, turbulence perturbations will decay and eventually die out.

However, the Boussinesq approximation has obvious problems. In the centre of a pipe with turbulent flow the mean flow has a zero gradient, $du/dr = 0$. The B.A. then predicts zero turbulent viscosity, yet from measurements we know that turbulence is most intense there. Clearly, approximating the turbulent Reynolds stress by a mean flow quantity cannot be correct.

The approximation for the Reynolds stress tensor in the Boussinesq model is

$$\tau_{ij} = -\rho\overline{u_i' u_j'} = \mu_T \left(\frac{\partial \overline{u_i}}{\partial x_j} + \frac{\partial \overline{u_j}}{\partial x_i}\right) \tag{6.26}$$

where the indices i and j run over the three spatial directions. This form is very similar to the laminar shear stresses, except that the laminar viscosity μ is replaced by a *eddy viscosity* μ_T.

6.6.1 Mixing length model

A very simple and early model is due to Prandtl. Although it is as simple as it can get, the model is surprisingly successful in predicting simple attached turbulent flows. Modified versions by *Baldwin-Lomax* or *Cebeci-Smith* are still widely used in structured grid methods for aeronautical flows. These models are also called *0-equation* models because no additional transport equation need to be solved for any turbulent quantities.

Prandtl observed the eddies when smoke billows from a smokestack, driven by the buoyancy of hotter, lighter smoke. A short distance from the outlet the strong differences in density are blended together, the buoyancy that drives the flow relents. To formulate a model, he surmised that given some length of travel, two eddies of disparate size will average to become of similar size. This is called the *mixing length hypothesis*:

$$|u'| \approx |v'| \approx l \left| \frac{d\bar{u}}{dy} \right| \tag{6.27}$$

where l is the *mixing length* and leads to

$$-\overline{u'v'} = l^2 \left(\frac{d\bar{u}}{dy} \right)^2 . \tag{6.28}$$

The fact that now $\frac{d\bar{u}}{dy}$ has been factored out of μ_T makes it possible to take very reasonable assumptions on the value of l in certain parts of the flow. For example, l can be taken to be proportional to the distance near the wall in regions close to the wall. This reflects the fact that larger scale eddies are damped by the proximity of the wall.

The mixing length hypothesis is rather crude and it is often difficult to choose a good length scale. Mixing length theory has, however, led to some very valuable results and is often quoted.

6.6.2 The Spalart-Allmaras model

An alternative approach to the simple mixing length model is to write a transport equation for a turbulent dissipation variable $\tilde{\nu}$. The transport allows to include non-local effects in the model, e.g., turbulence being generated at a particular spot and then transported downstream by the flow.

The derivation of the equations is beyond the scope of these notes. Just a verbal sketch of a general transport equation shall be given:

Rate of change of ϕ	+	Transport of ϕ by convection	=	Transport of ϕ by diffusion	+	Rate of production of ϕ	+	Rate of destruction of ϕ

Here ϕ stands for any quantity to be transported by the flow; in the case of the Spalart-Allmaras model it is the turbulent variable $\tilde{\nu}$.

The turbulent dissipation ν_T is then obtained by multiplying the turbulence variable $\tilde{\nu}$ with an attenuation function f_v which models the damping of the turbulence near the wall.

An equation for $\tilde{\nu}$ cannot be derived by theoretical analysis because ν_T is not a physical quantity. It merely arises as a product of the Boussinesq approximation. To derive the model Spalart and Allmaras selected a small number of viscous flows which they wanted the model to represent accurately. The structure of the convection, production and dissipation terms and their constants then can be derived by dimensional analysis or matching with experiments.

Maybe somewhat surprisingly for a model not developed from physical principles, it performs extremely well and has become the model of choice for many aeronautical applications. For external flows such as flows over aircraft the accuracy is equivalent to or better than the k–ε model, at the price of only one additional equation rather than two. Also, the Spalart-Allmaras model is somewhat better conditioned and more stable than the $k-\varepsilon$ model: the better stability increases the rate of convergence. It is able to predict small separation bubbles under certain conditions.

Depending on the implementation the model incorporates wall-functions or not. Fluent's implementation senses the grid resolution and can be run either in "wall-function" mode with $30 \leq y^+ \leq 60$ or in low-Reynolds mode with $y^+ \approx 1 - 2$.

It is less well suited for internal flows, such as flows in pipes, inlet ducts, etc. In these cases the variants of the $k-\varepsilon$ are better choices. The S-A model is not appropriate for flows with free turbulence such as thermal buoyancy. With an appropriate modification of wall distance that the model uses it can be used as a sub-grid scale model in a Large Eddy Simulation, a model termed the *Detached Eddy Simulation* or *DES*.

6.6.3 The $k-\varepsilon$ model

The most widely used turbulence model in CFD is the $k-\varepsilon$ model. Similar to the Spalart-Allmaras model it uses transport equations to include non-local effects. Differently from the Spalart-Allmaras model, not one but two additional transport equations are solved. Unlike the artificial turbulent viscosity variable $\tilde{\nu}$ of the Spalart-Allmaras mode, the transported turbulent quantities of the $k-\varepsilon$ model have physical meaning. The first variable is the turbulent kinetic energy k

$$k = \frac{1}{2}\left(\overline{u'u'} + \overline{v'v'} + \overline{w'w'}\right) \tag{6.29}$$

which can e.g., be measured in a wind tunnel from the turbulent fluctuations. The second variable is the viscous dissipation rate ε which governs the dissipation of turbulent kinetic energy due to the shearing of the smallest eddies measured as the fluctuation of the deformation e'_{ij},

$$\varepsilon = 2\nu\,\overline{e'_{ij}e'_{ij}}. \tag{6.30}$$

ε is given per unit mass and has dimensions $\mathrm{L}^2\mathrm{T}^{-3}$. To define an eddy viscosity in a mixing length model one needs a velocity scale and a length scale. A velocity scale can conveniently be taken from $u_{ref}\sqrt{k}$; similarly, a length scale is obtained from $l = k^{3/2}/\varepsilon$. The eddy viscosity then becomes

$$\mu_T = \rho C_\mu l u_{ref} = \rho C_\mu \frac{k^2}{\varepsilon}. \tag{6.31}$$

The structure of the terms in the transport equations can be derived on theoretical grounds. The non-dimensional coefficients which scale the terms have to be calibrated by comparison with experiments or simple cases where the exact solution is known. The derivation is beyond the scope of this book, but can be found in the literature.

The $k-\varepsilon$ model relies on the validity of the log-law for the flow. That is, there are two key areas where the model does not apply. One area is wall-bounded shear flows where there is no proper boundary layer such as stagnation or separation points. The other area is the viscous sub-layer below the log-law layer. The "standard" k–ε models bridge this region using appropriate boundary conditions for k and ε. This requires, however, that the first mesh point is located the within the log-law region; hence $30 \leq y^+ \leq 500$.

The boundary conditions for the wall shear stress exploit the fact that the overall shear stress in the inner layers is constant (cf. Sec. 6.4.2.2). We can, hence approximate the shear stress at the wall from the shear stress in the log layer. This approach is known as the *wall-function* approach.

If the first mesh point is below the log-layer but in the buffer region with $y^+ < 30$, the velocity profile has deviated from the log-law and the wall shear stress will be predicted wrongly. Similarly, if y^+ falls into the upper regions of the log layer, the log-law may not be applicable due to effects of pressure gradient. Practical requirements of mesh generation often force the user to compromise on satisfying these requirements.

In cases with strong pressure gradients or where separation occurs the damping functions for the linear sub-layer are inaccurate and the turbulence models need to resolve into the linear sub-layer. For more accurate solutions the $k-\varepsilon$ model can be coupled with a *low-Reynolds* model that is valid for the viscous sub-layer. Fluent e.g., uses the model by Wolfstein and calls the coupled model the "enhanced" $k-\varepsilon$ model. In this case the viscous sub-layer needs to be resolved with a number of mesh points. The first point should be located at $y^+ \approx 1-2$.

6.6.3.1 Advanced turbulence models: realisable k–ε, RNG, second-order closure

Turbulence modelling is at times a rather crude process and there is a lot of black magic involved in getting accurate CFD predictions for turbulent flows. However, for specific flows such as attached flows over aerofoils, the accuracy of the simple models is very good In general, though, the user will always have to validate his choice of model for a given test case against experimental evidence of similar flows.

Commercial codes offer a large variety of turbulence models for the user to choose from. There are around 5 popular dialects of $k-\varepsilon$ which may or may not perform better for a given application. The *realisable $k-\varepsilon$ model* modifies some of the constants in order to ensure that there actually is a solution to the equations. This might be a good model to try in cases where the standard $k-\varepsilon$ does not converge. The *renormalisation group model* (RNG) derives some of its structure of the $k-\varepsilon$ from statistical analysis considering the smallest scales. The $k-\omega$ model, a two-equation model for k and the specific dissipation rate ω, removes some of the problems with numerical stability related to the wall boundary condition on ε, but does not perform as well at the outer edge of the boundary layer or in free shear layers. The $k-\omega$ SST-model combines $k-\varepsilon$ in the outer regions and $k-\omega$ model near the wall, avoiding their weak points. While one of the best two-equation RANS models around, the switching between the models makes the stability of the SST model even more delicate than the stability of its component models.

A step up in cost and accuracy are second-order closure models or RSM which derive transport equations for the first-order correlations $\overline{u_i'u_j'}$. The arising triple correlations then have to be modelled. These models promise superior accuracy for strongly separated flows, but their lack of stability and robustness make them currently not suitable for general industrial use.

6.7 Near-wall mesh requirements

Section 6.4.2 has explained the different layers near the wall within the boundary layer, in particular the linear sublayer and the log layer. Exploiting the particular characteristics of these two layers we have two principal choices of modelling the behaviour of the flow near the wall, namely the low-Reynolds and the high-Reynolds approaches.

6.7.0.2 Low-Reynolds approach: resolution of the sublayer

In the low-Reynolds approach we discretise the domain fine enough such that the first solution point next to the wall is well within the linear sublayer with $1 \leq y^+ \leq 2$. The flow in this region can be considered laminar and we can use $\tau = \mu \frac{\partial u}{\partial y}$ to compute the wall shear stress τ. This condition should be satisfied for all areas where the wall shear stress plays an important role, but may be difficult to achieve in stagnation points.

This is a very expensive approach as the requirement of a very fine mesh near the wall will dramatically increase the number of cells in the mesh and hence the computational cost. Resolving the very steep gradients near the wall will also slow down convergence. However, if the assumption of constant shear stress in the inner layers is valid, there may not be a need to resolve this layer, as shown in the next section.

6.7.0.3 High-Reynolds approach: wall-functions

In the high-Reynolds approach we make use of *wall-functions* to estimate the wall shear stress by placing the first mesh point well within the log region, $30 < y^+ \leq 500$.[8] That is, the mesh is typically about 50 times coarser near the wall compared to the low-Reynolds approach.

The wall-function then estimates the wall shear stress by mapping the actual velocity gradient onto the gradient of the log-law and exploiting the assumption that the total wall shear stress $\tau = \tau_\mu + \tau_T$ is constant in the inner layers for zero pressure gradient. This approach typically works well for fully developed boundary layers with small or moderate pressure gradients.

In both cases we need to generate the mesh to have the appropriate mesh width normal to the wall:

1. Before generating the mesh we need to predict the wall shear stress for the flow using some approximation as discussed in the next section.
2. After an initial calculation, we need to then check the solution and verify that y^+ is indeed in the correct range. All CFD codes will offer post-processing utilities that allow to compute and visualise the y^+ distribution over the surfaces.

6.7.1 Estimating the wall distance of the first point

As discussed in Sec. 6.6.3, unless the implementation specifically allows to include the buffer layer $5 \leq y^+ \leq 30$, we should avoid as much as possible to have y^+ drop below 30 as the log-law is not valid there. In practice it will be complex and cumbersome to specify a varying physical wall distance y for the mesh generation such that in the final solution with varying τ_0 we obtain a y^+ in the correct band. For convenience we prefer to specify a constant value for y, and have to hence tolerate a variation in y^+. Using a lower target of $y^+ > 60$–100 will help to reduce the number of cells with $y^+ < 30$.

Similarly, limiting $y^+ \leq 100$–200 avoids the upper end of the log layer that may be affected by variation in pressure gradient, as at the top end of the log layer we are approaching the zone where the constant wall shear stress assumption is valid and the profile becomes more susceptible to pressure gradients.

Once we have selected a target y^+ value, we can design the mesh. The first step is to estimate the wall shear stress τ_0 which in turn allows to compute the physical wall distance of the first point y. Given a cell height outside the boundary layer that is appropriate for the mainly inviscid flow there, we can then determine growth rate and number of cells.

[8]When designing your mesh you will want to avoid placing your first mesh points at the extreme ends of the bracket. See Sec. 6.7.1.

6.7.1.1 Estimating skin friction for mesh generation

Given a target value of y^+ appropriate for the chosen near-wall modelling approach, we can compute the physical cell height y of the first cell by 'unscaling' $y = y^+\sqrt{\rho/\tau_0}$. This in turn needs an estimate of the wall shear stress τ_0.

A widely used approach to approximate turbulent boundary layers has been suggested by Prandtl. He viewed the turbulent flow in pipes as very similar to turbulent boundary layers. Especially in the near-wall regions, the flows should behave very similarly. In a slightly crude approximation one can equate the radius of a pipe with the boundary layer thickness and the maximum velocity in a pipe corresponds to the freestream velocity of a boundary layer. Prandtl's analysis starts with assuming that the wall shear stress for a boundary layer $|\tau_0|$ is given by Blasius' formula for the turbulent pipe with $Re < 10^5$,

$$|\tau_0| = \frac{1}{2}\rho\bar{u}^2 0.079\left(\frac{\nu}{\bar{u}2R}\right)^{\frac{1}{4}}. \tag{6.32}$$

He then proceeds to replace references to the pipe radius R with the boundary layer thickness δ. When considering the momentum balance one has to replace the pressure gradient that drives a pipe flow with the momentum lost due to the increased thickness of the boundary layer. The derivation will not be given here; it can be found e.g., in White [6].

Blasius' approximation is only valid for moderate Re, and in that range experiments determine the skin friction coefficient c_F for a turbulent boundary layer without pressure gradient to

$$c_F = 0.074(\mathrm{Re}_x)^{-\frac{1}{5}}, \qquad \mathrm{Re}_x < 10^7. \tag{6.33}$$

Schlichting[9] found for higher Reynolds numbers

$$c_F = \frac{0.455}{(log_{10}(\mathrm{Re}_x))^{2.58}}, \qquad 10^7 \leq \mathrm{Re}_x \leq 10^9. \tag{6.34}$$

The behaviour of c_F can be seen in Fig. 6.8.

As an example for the y^+ calculation, let us consider a flat plate in air with $\nu = 1.5\cdot 10^{-5}\,\mathrm{m}^2/\mathrm{s}$, $U_{Ref} = 15\,\mathrm{m/s}$, $\mathrm{Re} = 1\cdot 10^6$, length of plate $L = 1\,\mathrm{m}$, target $y^+ = 30$. The skin friction coefficient becomes

$$c_f \approx 0.074\,\mathrm{Re}_D^{-0.2} = 0.00467,$$

which leads to a friction velocity of

$$u_\tau = U_{Ref}\sqrt{\frac{c_f}{2}} = 0.725\,\frac{\mathrm{m}}{\mathrm{s}}.$$

The first cell height can then be determined as

$$y = y^+_{Req.}\frac{\nu}{u_\tau} = 30\frac{15\cdot 10^{-6}\frac{\mathrm{m}^2}{\mathrm{s}}}{0.725\frac{\mathrm{m}}{\mathrm{s}}} = 0.621\,\mathrm{mm}.$$

[9] Hermann Schlichting, German fluid dynamicist, 1907–1982.

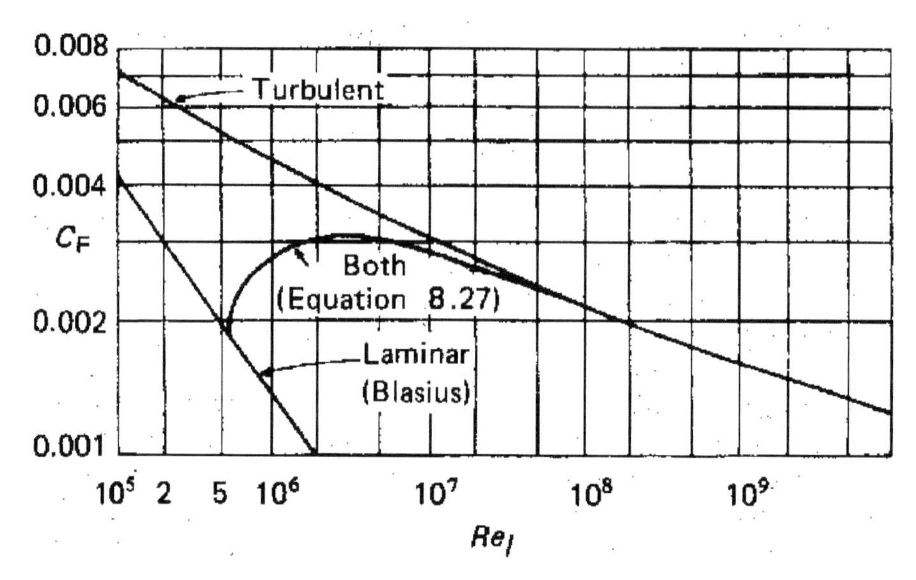

FIGURE 6.8
c_F for a turbulent boundary layer from White [6].

6.7.1.2 Number of points in the boundary layer, growth ratio

The velocity profile in the boundary layer exhibits very rapid change near the wall. The large gradients will lead to a large truncation error which needs to be balanced with appropriately small mesh width normal to the wall. As a guideline, if wall-functions are used (high-Reynolds approach), 60–100 $< y^+ \leq$ 100–200, which avoids to discretise the innermost layers with the steepest gradients, a minimum of about 10 points need to fall within the boundary layer thickness δ.

In the case of a low-Reynolds model with $1 \leq y^+ \leq 2$ we will need at least around 10 mesh points to resolve the inner layers up to $y^+ = 30$, resulting in 20-30 points in the boundary layer overall.

Blasius' approximation can also be used to derive an approximation to the boundary layer thickness,

$$\delta(x) = 0.381 \left(\frac{\nu}{u_\infty} \right)^{\frac{1}{5}} x^{\frac{4}{5}}, \tag{6.35}$$

where the constant 0.381 is determined by matching experiments. For the above formula it has been assumed that the turbulent boundary layer starts at $x = 0$, i.e., the length of laminar boundary layer before transition is considered negligible.

The flow outside the boundary layer will typically exhibit very small viscous effects, and the mesh will be designed primarily to correctly capture changes in the convective field. These consideration will determine the mesh width just outside the boundary layer. To maintain a smooth mesh without large jumps in cell size, we want the boundary layer mesh to smoothly blend into the remaining mesh with comparable cell widths.

If we fix the *growth ratio* or *stretching*, i.e., the ratio of cell heights in neighbouring cells $\sigma = h_{n+1}/h_n$, we can determine the number of cells N that are needed to grow from $y_1 = h_1$ to the height of the final cell h_N. Similarly, if we fix the number of cells to grow from $y_1 = h_1$ to h_N, this determines the value of σ. With σ, y_1 and N we can work out how many cells fall in the innermost layer below $y^+ = 30$ or in the boundary layer.

Let us consider again the example of the flat plate of the preceding section, $\mathrm{Re} = 1{\cdot}10^6$, length of plate $L = 1,\mathrm{m}$. The boundary layer height at L becomes

$$\delta = L \cdot 0.376 \left(Re_x\right)^{-1/5} = 1\,\mathrm{m} \cdot 0.376 \left(10^6\right)^{-1/5} = 0.0237\,\mathrm{m}.$$

If the stretching σ is to be 1.15 (a rather stringent quality value), and there are to be 10 cells ($N = 10$) in the boundary layer, let us estimate the distance of the first grid line from the wall y_1^+.

Summing up the cell heights normal to the wall produces

$$\delta = y_1 + \sigma y_1 + \sigma^2 y_1 + ...\sigma^{N-1} y_1 = y_1 \left(\frac{1-\sigma^N}{1-\sigma}\right).$$

In our example this gives

$$\frac{\delta}{y_1} = \frac{0.0237\,\mathrm{m}}{6.21 \cdot 10^{-4}\,\mathrm{m}} = 38.2 = \left(\frac{1-\sigma^N}{1-\sigma}\right).$$

The rational expression for σ on the right-hand side can be tested iteratively. In this example, in order to have 10 cells within the boundary layer we can have a stretching of at most $\sigma \leq 1.28$, or if we accept a higher computational cost and choose to have 15 points for more accuracy, $\sigma \leq 1.13$.

6.7.1.3 Summary of meshing for turbulent flows

In summary, the mesh generation for turbulent flows proceeds by the following steps:

1. Select the appropriate near-wall modelling approach. Either low-Reynolds with resolution of the linear sublayer and $1 \leq y^+ \leq 2$ (Sec. 6.7.0.2, or use of wall-functions with 60–100 $\leq y^+$ 100–200 (Sec. 6.7.0.3.
2. Estimate the wall shear stress τ_0 (Sec. 6.7.1.1) and boundary layer thickness δ (cf. Sec. 6.7.1.2).
3. Select a value of stretching σ and check whether there is a sufficient number of points N in the relevant layer: $N \geq 10$ in the boundary

layer for wall-functions or $N \geq 10$ below $y^+ = 30$ and $N \geq 20$–25 in the boundary layer for low-Reynolds.

Alternatively, fix the number of points and compute the resulting stretching σ which should not exceed 1.3.

4. Run the computation.
5. Verify in the actual results whether the initial approximation to τ_0 and δ were correct. If not, adjust the mesh and recompute.

6.8 Exercises

6.1 What are the main characteristics of turbulent flow?

6.2 Explain the Kolmogorov cascade.

6.3 Describe the 'outer' scaling for a boundary layer

6.4 Describe the 'inner' scaling for a boundary layer and the different near-wall layers that can be distinguished.

6.5 Explain the differences between RANS, LES and DNS.

6.6 What are zero-, one- and two-equation RANS models? Give an example for each type.

6.7 Explain the two main approaches for near-wall modelling.

6.8 You are asked to generate a mesh for an aerofoil operating at Re=10^4 using wall-functions. Estimate the wall distance for the first node. You want to have 15 nodes in the boundary layer. Estimate the maximal stretching you are allowed.

7

Mesh quality and grid generation

The mesh and its characteristics have a significant influence on the accuracy of the solution, and its computational cost. Whether or not the user runs on a pre-produced mesh, or whether the user generates her/his own, a competent analysis of the accuracy and potential errors in the solution do require a sound understanding of how the grid affects the solution. This is presented in Sec. 7.1.

A competent user will generate meshes to suit the case. He/she needs to first select an appropriate mesh generation approach and then appropriately control the parameters of the mesh generation algorithm. Secs. 7.2-7.4 present the approaches that are currently most widely used.

An outline of methods to have the solver automatically adapt the mesh to a solution on an initial mesh is given in Sec. 7.5.

7.1 Influence of mesh quality on the accuracy

7.1.1 Maximum angle condition

One of the most classic results in the analysis of mesh quality is by Babuška and Aziz [7]. The reader shall be spared the mathematics and the notion of Sobolev spaces, but, in brief, it states that if the maximum angle in a cell approaches 180°, the interpolation error becomes unbounded. This has two implications for mesh generation. Firstly, when generating isotropic meshes (outside of boundary layers) the cells ought to be as equiangular as possible. The smaller the maximum angle, the better; a good mesh will have orthogonal quadrilateral cells and equilateral triangles.

Equiangular triangles are not always possible. The size of the discretisation molecule in each direction should match the gradients of the solution since the truncation error is proportional to the first or second derivative of the solution, depending on the discretisation and the chosen accuracy of the scheme.

In boundary layers, e.g., the cells need to be much smaller in the direction perpendicular to the wall compared to along the wall, since gradients normal to the wall are very large, but gradients tangential to the wall are much smaller. In an attached boundary layer the ratio of the gradients is $\approx \sqrt{\mathrm{Re}}$. Aspect ratios of over 1000 in high-Reynolds number boundary layers are not

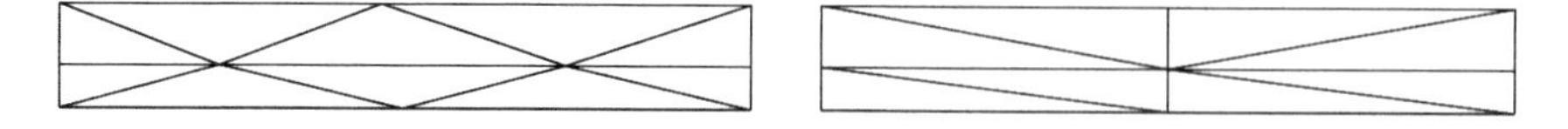

FIGURE 7.1
Obtuse (top) and the preferable non-obtuse (bottom) triangulations of a stretched mesh.

uncommon. Figure 7.1 gives an example of a triangulated stretched mesh. The triangulation needs to be carefully chosen to avoid excessive angles. A better solution yet is not to use a triangular or tetrahedral mesh where high anisotropy is required, but to use the hexahedral or prismatic cells that can be "squashed" normal to the wall without affecting grid orthogonality (cf. Section 7.4.5).

In Fig. 7.2 one can observe how poor grid quality affects the solution. The flow is entering on the left before interacting with a cavity, and then exiting on the right. As the flow enters fully developed, one would expect perturbed pressure contours in the cavity, but in the inflow section, where the flow is fully developed, the pressure contours should be parallel, exhibiting a pressure gradient only in the axial direction. The obtuse cells in that part of the domain

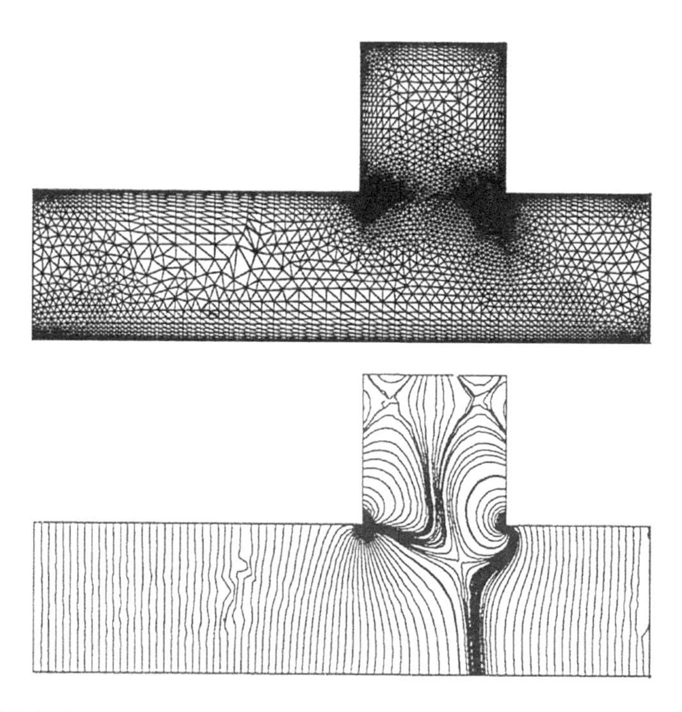

FIGURE 7.2
Grid for channel flow with cavity. Obtuse elements (top) link to perturbations in the pressure field (bottom).

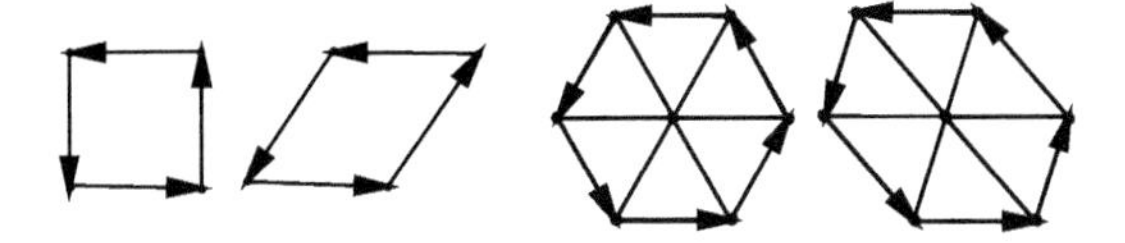

FIGURE 7.3
A few regular n-gons with n=4,6 that lead to a second-order accurate residual calculation with Cell-Vertex schemes.

induce a strong pressure perturbation. We should aim to have element angles below 120°–130°. How much poor angularity affects the solution does depend on the particular discretisation.

7.1.2 Regularity

A large class of unstructured finite volume methods is only fully second-order accurate if the cells connected to a node form a regular n-gon e.g., with $n = 4$ or $n = 6$ in 2-D. Regular means here that the domain can be tiled with cells of the same shape. In practice this means for triangulations that around each node one desires to have six triangles with parallel bases on opposite triangles (Fig. 7.3).

Note that as opposed to triangles in two dimensions, tetrahedra cannot fill three-dimensional space in a regular fashion. A better choice is to use regularly placed hexahedra where possible. In 3-D grids prismatic cells are often used in regions where high stretching is required for boundary layers. The surface triangulation then needs to be as regular as possible to have a regular honeycomb pattern for the tangential two-dimensional plane, while the normal plane has regular quadrilateral faces.

We can observe the effect of grid irregularity in Fig. 7.5. The mesh is generated by triangulating the circular cross sections at the vessel ends; these triangulations are then extruded to produce layers of prisms along the vessel axes. The grid is regular in the axial direction with quadrilateral faces due to the regular extrusion, Fig. 7.4 (b). Regularity in the cross section is provided by the triangulation having many vertices with 6 triangles around it producing a regular honeycomb shape.

The computed wall shear stress (WSS) shown in Fig. 7.5 jumps up by nearly 50% when the mesh transitions from prisms to tetrahedra. This is physically incorrect; the error is due to the irregularity of the mesh.

Better discretisations can help reduce these effects. The simulation of Fig. 7.5 was performed with Fluent using the standard discretisation of gradients. If the user selects the more expensive "node-based" gradient evaluation, the error is significantly reduced.

One could construct a multi-block mesh with very high cell regularity and orthogonality for this geometry; however, the block setup would be quite complicated and would take days to produce. What has been done here is

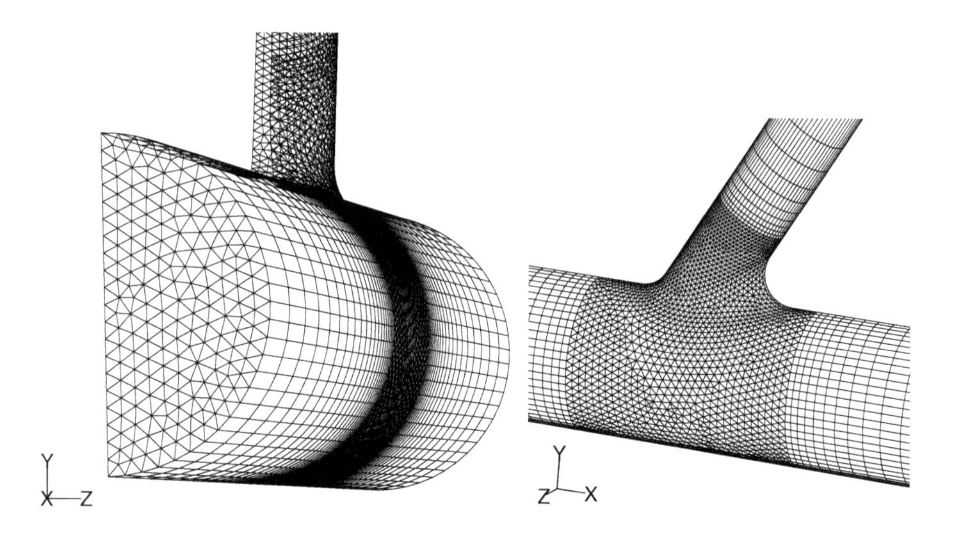

FIGURE 7.4
Hybrid mesh for a blood vessel bifurcation. Left: triangular meshing in a two-dimensional cross section. Right: extrusion to form linear segments with prism elements, with a tetrahedral mesh in the bifurcation zone.

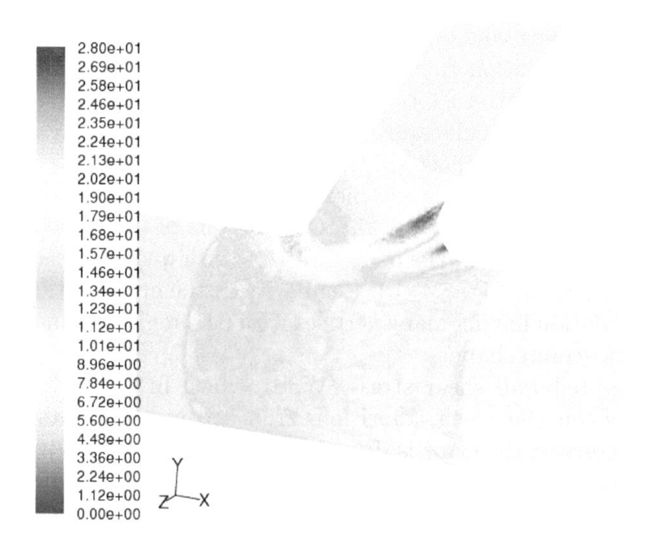

FIGURE 7.5
Flow in a bifurcation. The predicted value of wall shear stress dramatically increases due to the irregularity in the tetrahedral mesh zone.

the alternative of using a very simple mesh generation approach, taking only minutes to produce, but paying for that convenience with much larger errors in the simulation. This in turn can be in many instances compensated by using a finer mesh, hence trading relatively inexpensive computer time against very expensive user time.

Very often with mesh generation it is "you get, what you pay for" and the user has to judge what level of effort for the mesh generation is justified by the required accuracy.

7.1.3 Size variation

An intuitive feel for the effect of size variation can be derived by looking at a simple finite difference formula, a forward difference of a first derivative:

$$u_x = \frac{u_{i+1} - u_i}{h}. \tag{7.1}$$

This gradient approximation has been shown in Eq. 4.1 as being first-order accurate. However, performing a Taylor series not around x_i, but around the midpoint $\frac{1}{2}(x_i + x_{i+1})$, one finds it is second-order accurate there. In fact, the one-sided difference is a central difference for the midpoint.

Hence, to maximise the accuracy of the flux approximation, the cell interface should be at the midpoint between both nodes $i, i+1$. In graded fields where the mesh size varies one has to accept variations in cell size. The ratio between successive cell sizes is often referred to as *stretching*. The differences ought to be, however, as small as possible. In practice stretching of 1.1-1.2 is fully acceptable, but it should be kept below 1.3 in regions where important flow features need to be captured. The stretching can be allowed to take very large values in regions where the flowfield is near-uniform, e.g., in far-field regions for external aerodynamics simulations.

7.2 Requirements for the ideal mesh generator

Mesh generation is a crucial part of CFD. It has a significant influence on the runtime and memory use of the simulation, as well as the accuracy and stability of the solution. To start with, one could formulate a list of characteristics of the 'ideal' mesh generator:

Simple to use: It uses minimum user input; even the novice can produce a consistent grid for a first solution, while still allowing the expert to tailor the grid to his needs. This means that user-time should be small.

Fast to generate: It uses small amounts of memory and CPU time to generate the mesh.

Good boundary discretisation: It should be able to represent the boundaries accurately; finally it is the boundary conditions that define the problem. This should, of course, be possible for any kind of geometry.

Good interior discretisation: The mesh quality should be high: the grid should be as regular as possible and the element size should vary smoothly and in accordance with the sizes desired by the user. Elements should not exhibit large angles.

Control over the mesh size: It should be possible to easily and locally control the mesh size in the various regions of the domain.

Anisotropy: For many problems it is important to specify non-isotropic behaviour, e.g., highly stretched cells to capture boundary layers. The mesh generator should allow to produce high-quality high-aspect ratio cells.

Fast to compute on: With the same number of points in a mesh, the time to converge on the mesh can vary significantly for different types of meshes. Hexahedral meshes, as long as the elements are well shaped with orthgonal sides, typically incur lower computational cost, so hexahedral meshes should be preferred.

Today, there exists no mesh generation algorithm that would satisfy all of the criteria above. Many techniques are being used; all of them are still being developed further. The major techniques that are widely used today will be briefly presented in the following.

Two major categories of grid types are commonly distinguished: structured grids formed of regular lattices and unstructured grids formed of arbitrary collections of elements. Recently hybrid grids have become very popular that combine structured and unstructured parts. All of the following discussions will be presented in two dimensions, since this is easier to visualise but it still presents the major characteristics of the various approaches. Where relevant, their extension to three dimensions is also discussed.

7.3 Structured grids

Structured grids assume that the grid is topologically a rectangular array (Fig. 7.6). The grid is described to the solver by the dimensions of the array, (m, n), and a list of coordinates for the grid-points, in the following, called the nodes. A cell (i, j) is then formed by the 4 nodes (i, j), $(i+1, j)$, $(i, j+1)$ and $(i+1, j+1)$.

All early development of CFD was done using structured grids and solvers that made convenient use of the rectangular topology. In a structured grid the neighbours of a cell can immediately be calculated: the neighbour to the "right" of cell i, j is cell $i+1, j$. In many incompressible finite volume methods the control volume is hence taken as the cell, the *cell-centred* approach, which we have already used in Sec. 3.4.1.

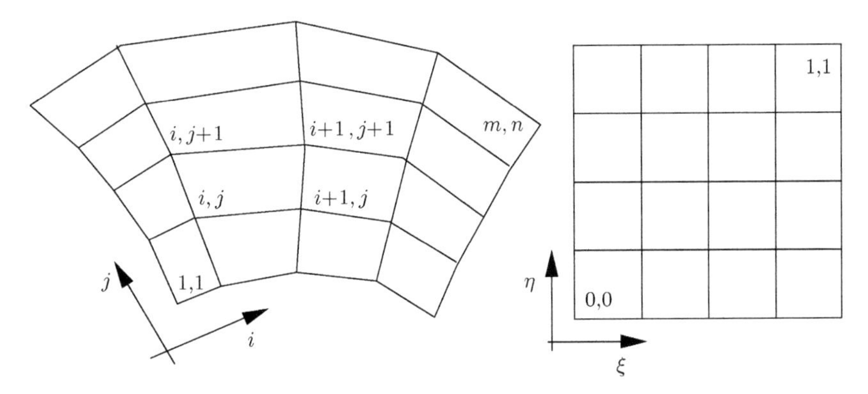

FIGURE 7.6
A structured grid is topologically rectangular.

Alternatively one could view the finite volume (i, j) as the volume around node (i, j), formed by connecting the centres of the 'cells' around to node (i, j). This *node-centred* or *vertex-centred* approach is often used for finite volume discretisations of the compressible equations on unstructured grids (cf. Sec. 7.4). Regardless of where one locates the unknowns in the grid, the count of nodes and computational cells remains the same in a quadrilateral or, similarly, a hexahedral grid if the number of boundary nodes is negligible compared to the number of interior nodes,

$$\begin{aligned} m_{nodes} &= m_I \cdot m_J \\ m_{cells} &= (m_I - 1) \cdot (m_J - 1) \\ &= m_{Nodes} - m_I - m_J + 1 \approx m_{nodes}. \end{aligned}$$

As will be shown in Sec. 7.4, for unstructured meshes composed of triangles or tetrahedra there will be quite a difference whether we store at the cell centres or at the grid vertices.

7.3.1 Algebraic grids using transfinite interpolation

Transfinite interpolation is at the heart of most structured grid generation. Due to the difficulty of wrapping structured grids around complex geometries, they are currently used only in very few areas, such as high-fidelity wing-body simulations in aeronautics. However, structured grids around simple two-dimensional geometries are straightforward to generate and by precisely controlling the quality of the grid we can learn a lot about how sensitive the solution is to grid quality. This is demonstrated in the case study in Sec. 9.1.

The basic idea in algebraic grid generation is to interpolate on a set of curves in physical space one curve for each grid dimension as shown in Figs. 7.7 and 7.8. The user selects a distribution of points along the boundaries with a

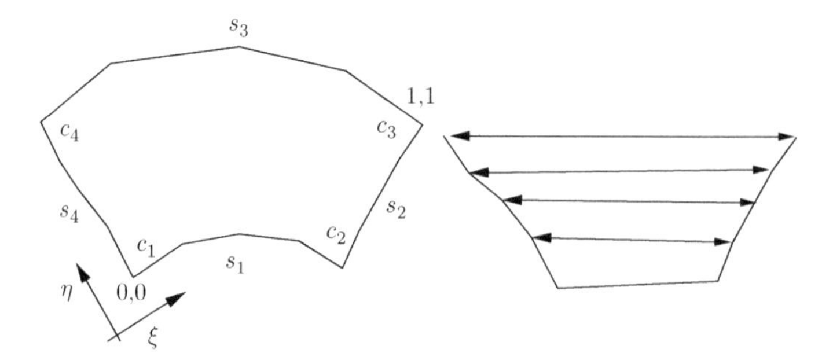

FIGURE 7.7
Interpolation of a grid along four sides $s_{1..4}$. The picture on the right shows one-dimensional interpolation along ξ.

matching number of points on opposite sides. The curves along the sides are denoted $\mathbf{s}_i$:

$$\mathbf{s}_i(\gamma), \qquad i = 1, 4; \quad 0 \leq \gamma \leq 1,$$

where γ is the arc-length along the side. Let us use the convention that positive arc length traverses the $\mathbf{s}_1, \mathbf{s}_3$ sides in the positive ξ and $\mathbf{s}_2, \mathbf{s}_4$ sides in the positive η direction. The sides form a topological rectangle with corners $\mathbf{c}_i$, $i = 1, 4$. As sketched in Fig. 7.7, the sides $\mathbf{s}_1$ and $\mathbf{s}_4$ join at corner $\mathbf{c}_1$,

$$\mathbf{c}_1 = \mathbf{s}_4(0) = \mathbf{s}_1(0),$$

and similarly for the other corners.

To find the interior mesh lines an interpolation function $\mathbf{g}(\xi, \eta)$ is needed which recovers the boundary curves. For example, along side $\mathbf{s}_1$, $\eta = 0$, one needs

$$\mathbf{g}(\xi, 0) = \mathbf{s}_1(\xi)$$

and similarly for the other sides. One-dimensional interpolation between opposite curves (Fig. 7.7) gives

$$\mathbf{A}(\xi, \eta) = (1 - \xi)\mathbf{s}_4(\eta) + \xi \mathbf{s}_2(\eta)$$
$$\mathbf{B}(\xi, \eta) = (1 - \eta)\mathbf{s}_1(\xi) + \eta \mathbf{s}_3(\xi).$$

In a rectangular mesh with perpendicular ξ and η each interpolation is independent and the final interpolation is just $\mathbf{A} + \mathbf{B}$. For a curved boundary the sum of $\mathbf{A} + \mathbf{B}$ gives, e.g. for side $\mathbf{s}_1$.

$$\begin{aligned}(\mathbf{A} + \mathbf{B})(\xi, 0) &= (1 - \xi)\mathbf{s}_4(0) + \xi \mathbf{s}_2(0) + (1 - 0)\mathbf{s}_1(\xi) \\ &= (1 - \xi)\mathbf{c}_1 + \xi \mathbf{c}_2 + \mathbf{s}_1(\xi) \\ &\neq \mathbf{s}_1(\xi),\end{aligned}$$

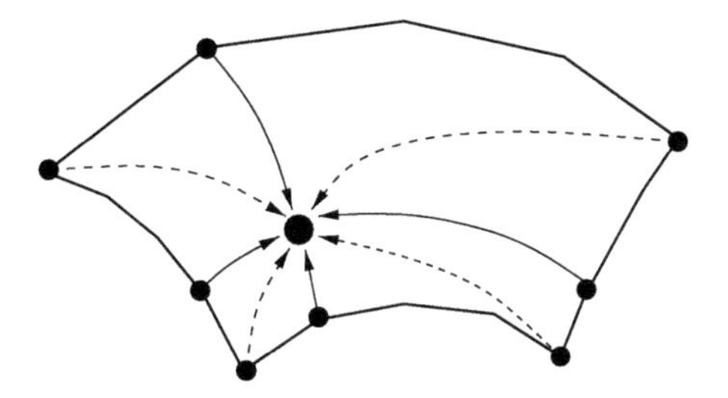

FIGURE 7.8
In transfinite interpolation the position of a node depends on the locations of the corresponding nodes on the sides (solid arrows) and on the location of the corners of the mesh (dashed arrows).

similarly for the other sides. This error can be removed from $\mathbf{A} + \mathbf{B}$ e.g., by adding for side $\mathbf{s}_1$,

$$-\frac{1}{2}(1-\eta)\left[(1-\xi)\mathbf{c}_1 + \xi\mathbf{c}_2\right].$$

The factor $(1-\eta)$ ensures that this correction becomes zero at the opposite side $\mathbf{s}_3$ where $\eta = 1$. The factor 1/2 comes about through symmetry. The extra terms appear at the connected sides $\mathbf{s}_2$ and $\mathbf{s}_4$ again.

Adding terms at all four corners, the interpolation function is then found as

$$\begin{aligned}\mathbf{g}(\xi,\eta) =& (1-\eta)\mathbf{s}_1(\xi) + \xi\mathbf{s}_2(\eta) + \eta\mathbf{s}_3(\xi) + (1-\xi)\mathbf{s}_4(\eta) \\ &- \left[(1-\xi)(1-\eta)\mathbf{c}_1 + \xi(1-\eta)\mathbf{c}_2 + \xi\eta\mathbf{c}_3 + (1-\xi)\eta\mathbf{c}_4\right].\end{aligned} \tag{7.2}$$

The interpolation $\mathbf{g}$ is called *transfinite interpolation.* Figure 7.8 shows the calculation of the location of an interior node i, j. It depends on the location of the boundary nodes of "column" i and "row" j, as well as on the location of the corner nodes. The calculation is inexpensive, since the location of every interior node can be calculated directly from (7.2). A solution of a linear system of equations is not required.

More sophisticated formulations are possible and lead to *elliptic grid generation* methods which need to solve a coupled set of partial differential equations for the node coordinates. More on elliptic meshing can be found in [8, 9].

Transfinite interpolation provides many control mechanisms to arrive at a desired point distribution, e.g., to achieve a dense mesh spacing in the η-direction to capture a boundary layer along side $\mathbf{s}_1$, it is sufficient to space the boundary nodes for low η on sides $\mathbf{s}_4$ and $\mathbf{s}_2$ accordingly (Fig. 7.9).

Also the angularity of the cells can be influenced by a suitable distribution of nodes along the boundaries. Fig. 7.10 shows an idealised mesh of a curved duct. On the left a poor distribution of the nodes on the outer boundary leads to very *skewed* cells with large corner angles. The improved boundary node distribution produces a smooth, nearly orthogonal mesh.

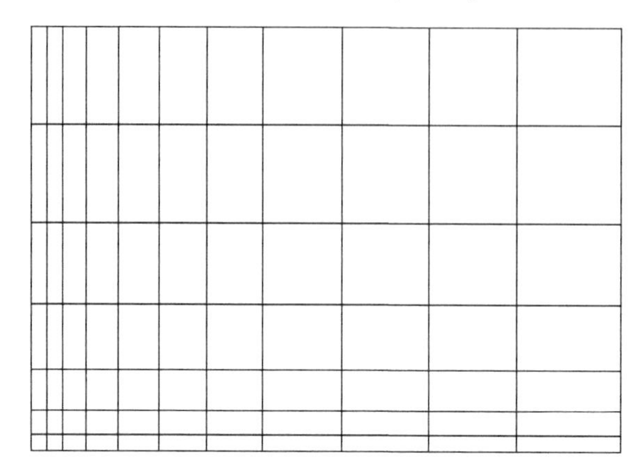

FIGURE 7.9
Boundary point control to cluster grid lines in transfinite interpolation.

For complex geometries smooth and high quality grids cannot always be achieved. They can even overlap and produce cells with negative volumes. One remedy is to use a grid that is composed of a number of *blocks* which join seamlessly, producing a *multi-block mesh*. The user needs to set up the block topology by specifying how the blocks connect and how many mesh points are placed along each block edge. At each block interface the user needs to specify the geometry of the interface curve and the point distribution on it. Better mesh quality can be also obtained with elliptic methods which allow to compute the best position and point distribution for the block interfaces.

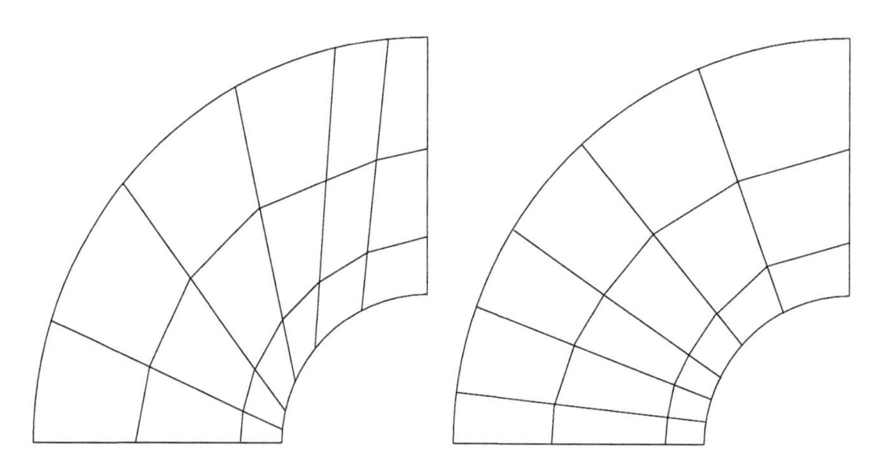

FIGURE 7.10
Boundary point control to improve grid angles in transfinite interpolation.

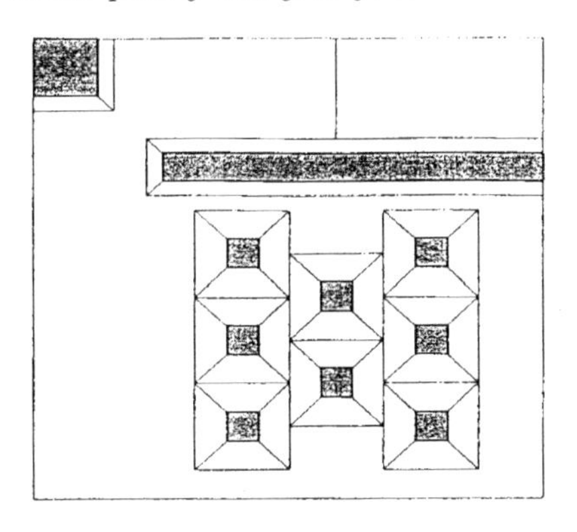

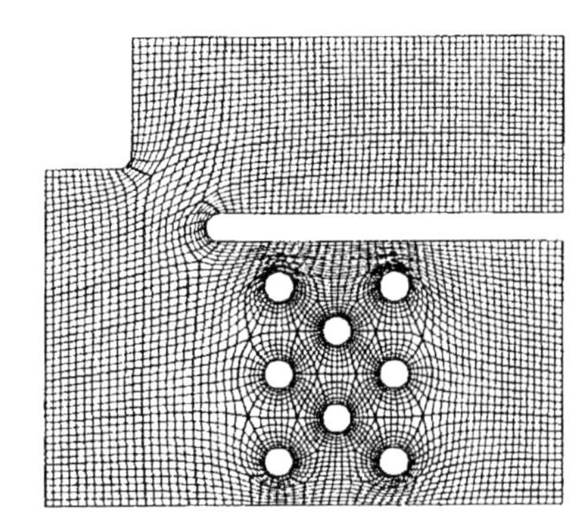

FIGURE 7.11
Multi-block mesh for a channel with a splitter plate and blocking pipes. Block topology on the left, multi-block mesh on the right.

Multi-block mesh generation can produce very high-quality grids. It is still used for high-accuracy high-Reynolds number simulation e.g., for turbo-machinery or aeronautical flows. However, the effort in designing the block topology is significant, requiring a lot of experience and skill by the operator, and it is hence no longer a widely practised approach.

7.4 Unstructured grids

Unstructured grids do not assume any fixed connectivity but rather describe the mesh as a collection of cells or *elements* that have a list of forming nodes or *vertices* as shown in Fig. 7.12. This list is called the *connectivity table*. While

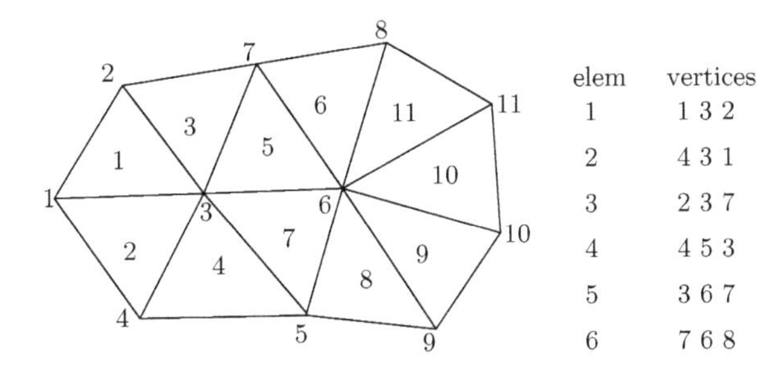

elem	vertices
1	1 3 2
2	4 3 1
3	2 3 7
4	4 5 3
5	3 6 7
6	7 6 8

FIGURE 7.12
An unstructured triangular grid is described by a connectivity table. Cell numbers are listed in the centres of the cells, node numbers near the nodes.

this description allows for a very flexible mesh that can follow any complex geometry, there is an added storage cost for the connectivity and an overhead for looking up the number of the forming nodes in the table before performing a flux calculation, a process called *indirect addressing*. Unstructured grids are not necessarily triangular or tetrahedral. For hybrid grids, e.g., in 2-D composed of triangles and quadrilaterals, one needs to carry extra information in the connectivity table which specifies the type of the cell.

All modern commercial CFD codes are hybrid unstructured, supporting a range of elements such as triangles and quadrilaterals in 2-D, tetrahedra, prisms, pyramids and hexahedra in 3-D. A structured 2-D grid can, of course, be viewed as an unstructured quadrilateral grid; the converse is not necessarily possible.

Instead of storing the elements and their forming nodes, it is actually much more efficient for the flow solver to store the face between two control volumes. Some faces of the grid in Fig. 7.12 are given in Table 7.1. To accumulate the flux balance for cell 7 we need to compute the fluxes across the faces 3, 8 and 9. Each face is only visited once since the flux leaving the cell on one side of the face has to be exactly the same flux as for the cell on the other side. Only the sign switches. In this example the orientation of faces 3 and 9 have the flux signed to enter cell 7, while face 8 is orientated the other way.

Listing faces rather than elements not only simplifies the calculation, it also allows to treat arbitrary polyhedral elements with more faces than the standard element types such as tetrahedra and hexahedra. General polyhedra may arise e.g., when a simple rectangular block of mesh is intersected with an oblique geometry, leading to cut cells.

The count m of edges, cells and nodes of a two-dimensional grid is given by Euler's formula

$$m_{cells} = m_{edges} + m_{nodes} + 2.$$

face no	node left	node right	cell from	cell to
1	1	3	2	1
2	3	4	2	4
3	3	5	4	7
...	...	...	...	...
8	6	5	7	8
9	6	3	5	7
...	...	...	...	...

TABLE 7.1
Unstructured two-dimensional grid connectivity given by list of faces.

To obtain relations between the number of nodes and the number of triangles or edges in a grid while neglecting boundary effects, consider a structured, or quadrilateral, grid that is cut into triangles. On the quadrilateral grid one finds (cf. Sec. 7.3) $m_{nodes} \approx m_{cells}$. By cutting each quadrilateral in two triangles, one obtains

$$m_{triangles} = 2m_{nodes}. \tag{7.3}$$

A two-dimensional triangular mesh has nearly twice the number of cells as it has nodes. A three-dimensional tetrahedral mesh has 5-6 times as many tetrahedra as nodes. That is, for the same number of nodes a tetrahedral grid requires 5-6 times the amount of memory as compared to a hexahedral grid.

Consequently, as opposed to quadrilateral or hexahedral grids, there is a significant difference whether the unknowns are placed at the nodes, as in *node-centred* or *vertex-centred* methods, or whether they are placed at the cell-centres as in *cell-centred* methods. Most incompressible unstructured solvers use the cell-centred approach (OpenFOAM, Fluent, Star-CCM+, ACE+), but there are also vertex-centred examples such as CFX from Ansys. Unstructured compressible codes, which often are in-house codes in the aeronautical industry or research codes by the government research labs NASA, DLR or ONERA, are predominantly vertex-centred. Examples are Rolls Royce's hydra code, DLR's tau code or NASA's fun3d code.

On the other hand, when counting the faces, the argument reverses. While storing at cell-centres on a two-dimensional triangular grid requires twice the storage, each triangular cell requires only three flux calculations across its three sides. The vertex-centred storage has 6 neighbours, but there is only half of them. Hence in terms of effort of flux computation, both approaches are equivalent. There is no clear overall advantage for either approach; the reason for incompressible codes favouring cell-centred storage is mainly historic.

Nevertheless, based on the same number of nodes hexahedral grids, e.g., obtained through structured grid generation, are much more efficient in storage and computational effort than unstructured tetrahedral grids using the same number of nodes. While this seems to strongly favour structured grid methods, unstructured grid methods provide a much greater flexibility in refining the mesh where needed and using larger cells in regions of lesser interest. In structured grid methods mesh lines need to be continued, which severely curtails the freedom in mesh design. Hybrid grid generation methods exist that combine the two approaches (cf. Sec. 7.4.5).

There are three major methods to generate unstructured tetrahedral meshes. These introductory notes cannot provide you with a detailed description, but shall attempt to give you an idea how they work and what their main properties are.

7.4.1 The Advancing Front Method

The *advancing front method* (AFM) starts in the two dimensional case with a discretisation of the boundary curves just as in algebraic structured meshing

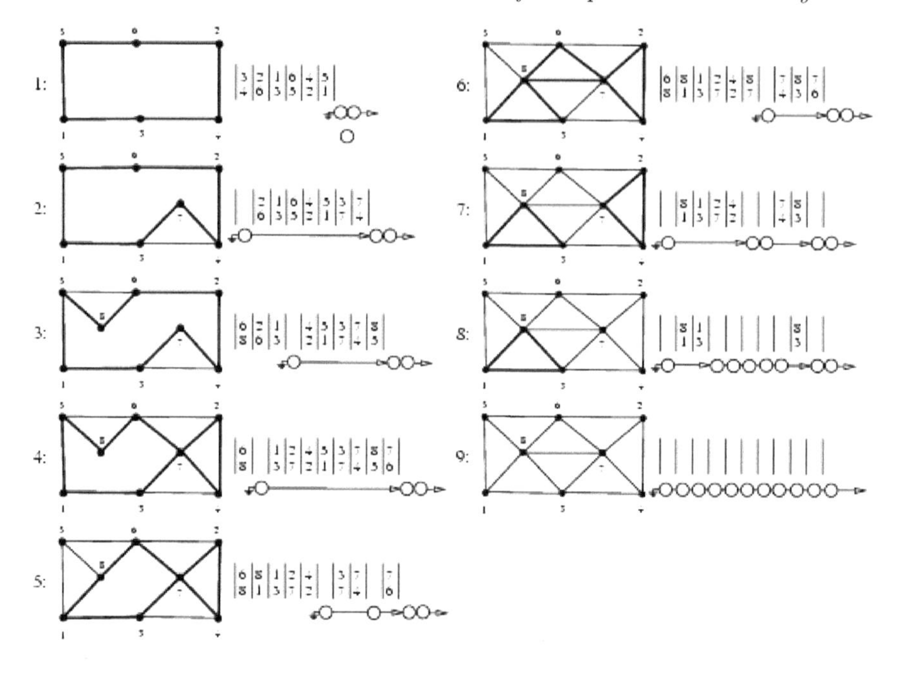

FIGURE 7.13
Sample steps of the advancing front method in 2-D. The front is given as the list of edges between two nodes on the right.

(cf. Sec. 7.3.1), forming a list of linked edges, the *front*, as shown in Fig. 7.13. One of these edges is picked and a well-shaped triangle is built upon it protruding into the domain. The constructed triangle is added to the list of elements of the mesh; the list of edges is updated from which the next edge is picked. The process terminates once the final "hole" in the domain has been meshed and no more edges remain in the front.

This type of algorithm is called *greedy* since each newly created triangle bites a chunk out of the domain. Once a triangle is formed it is never undone, which can lead to poor mesh quality in some instances or even failure to mesh the domain in others. In general, though, the mesh quality is quite good since triangles or tetrahedra are created in layers starting from the boundaries.

The extension to tetrahedra is straightforward, replacing the front of edges with a front of triangular faces and creating one tetrahedron at a time. The AFM is widely used, e.g., in the mesh generator CFD-GEOM of the commercial CFD package CFDRC.

7.4.2 Delaunay triangulation

Delaunay triangulation is a mathematical approach to *tessellation*, the tiling of a domain with triangles or tetrahedra. The most popular variant of the method

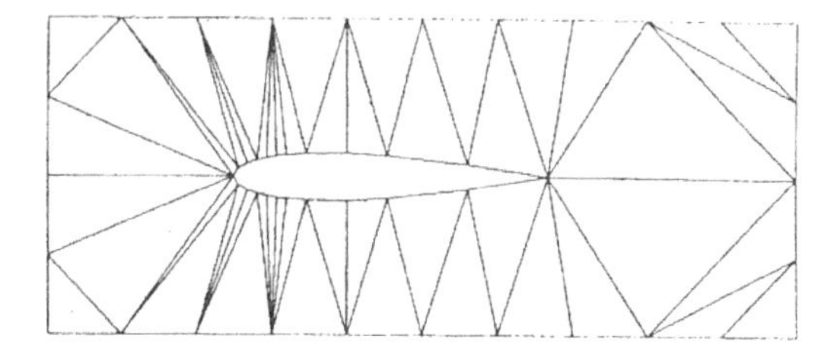

FIGURE 7.14
Initial triangulation of the boundary nodes.

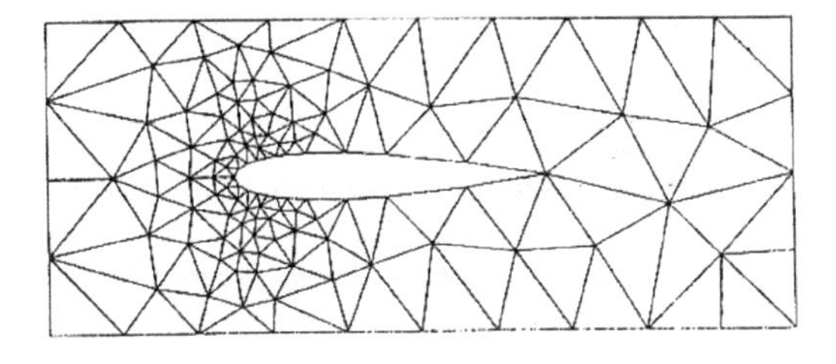

FIGURE 7.15
Final triangulation of the boundary nodes of Fig. 7.14.

works by insertion of one node after the other into an existing tessellation beginning with a single triangle or tetrahedron. A new vertex which is to be inserted will lie inside one triangle and also inside the circum-circles of a few more. Simple geometry shows that if one deletes all these triangles and connects the new vertex to all the exposed vertices of the resulting cavity, all the new triangles or tetrahedra will be non-overlapping and form a valid mesh.

Figure 7.14 shows a triangulation generated with this approach where all prescribed boundary nodes have been inserted. The next step is to add suitably chosen interior nodes until an appropriate mesh density is achieved (cf. Fig. 7.15). As can be seen from figures 7.14 and 7.15, the method is not greedy: the insertion of a new node will remove a number of existing triangles.

Fluent's unstructured mesh-generators TGrid and Ansys' ICEM, as well as the open-source mesh generator gmsh,[1] offer tetrahedral mesh generation through Delaunay triangulation. An example of a tetrahedral mesh generated by gmsh for an Airbus A319 aircraft is shown in Fig. 7.16.

7.4.3 Hierarchical grid Methods

Hierarchical methods were initially developed for grid generation and simulation in structural mechanics. The procedure is to perform regular Cartesian

[1] http://geuz.org/gmsh/.

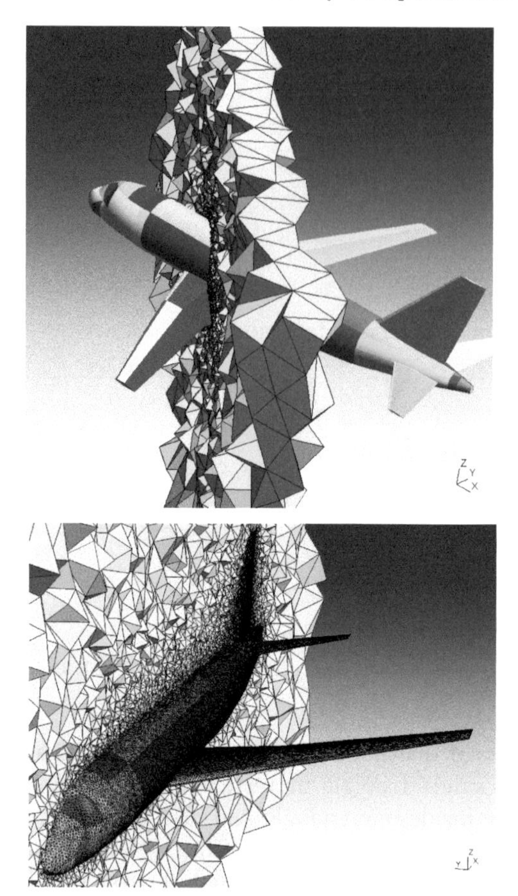

FIGURE 7.16
Tetrahedral grid around an aircraft created with the open-source tool gmsh. (Image courtesy gmsh.)

subdivisions of an initial square or cube box. The first division splits the *root* quad into 4 children (or a root cube into 8 children), each child again if required. Similarly one can work with triangles or tetrahedra instead of cubes; the subdivision remains the same. The final *tree* consists of a hierarchy of cells. The ones that form the computational mesh are the *leaves* which don't have children themselves. The jump in mesh size between neighbouring leaves is limited to a factor of two, i.e., one extra subdivision. The resulting meshes from quadrilaterals or cubes are often *triangulated*, as shown in Fig. 7.17.

A similar procedure can be formulated by recursively subdividing tetrahedra. The mesh generator TETRA produced by ICEM CFD is based on this subdivision.

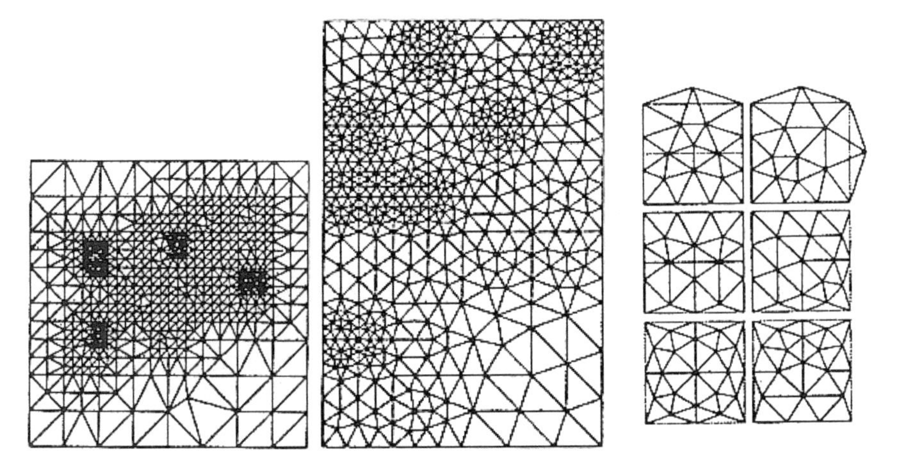

FIGURE 7.17
Resulting triangulation of a quad-tree method.

7.4.4 Hexahedral unstructured mesh generation

Hexahedral regular grids that are aligned with the flow present very good accuracy of the simulation at rather low computational cost. High-quality boundary-aligned hexahedral grids can only be achieved with structured multi-block mesh generation (cf. Sec. 7.3.1) which is to cumbersome to set up for complex geometries. A number of mesh generators recently have taken a shortcut approach, where a regular structured mesh is aligned with the coordinate directions x, y, z, hence termed a *Cartesian*[2] *mesh.*

This mesh is then hierarchically refined in zones of interest to achieve mesh gradation; refinement boxes are selected by the user. Fig. 7.18 (a) shows 4 levels of refinement around a vehicle body in a wind tunnel. The mesh will be irregular where two levels of refinement meet. Hexahedral cells with *hanging nodes* will be generated where a coarse level cell is adjacent to a fine level cell which has additional nodes at the mid-edge and mid-face position. If a polyhedral approach is taken this is not a problem. The element will simply be formed of more than 6 faces. In the example shown in Fig. 7.18 (a) the cells with hanging nodes have been subdivided into other elements such as tetrahedra and pyramids in order to avoid having to deal with the polyhedra.

The geometry will not perfectly align with the grid lines but cut into the mesh, resulting in a Lego-like or castellated shape where the mesh intersects the surface (Fig. 7.19 (a)). These cut cells can then be fixed by dividing or combining cells near the surface, resulting in a smooth, repaired mesh (Fig. 7.19 (b)).

The refinement of the mesh in blocks is not the most efficient approach. It is difficult to finely tailor the mesh density. On the other hand, using hexahedra

[2]René Descartes, French philosopher, mathematician, 1596–1650.

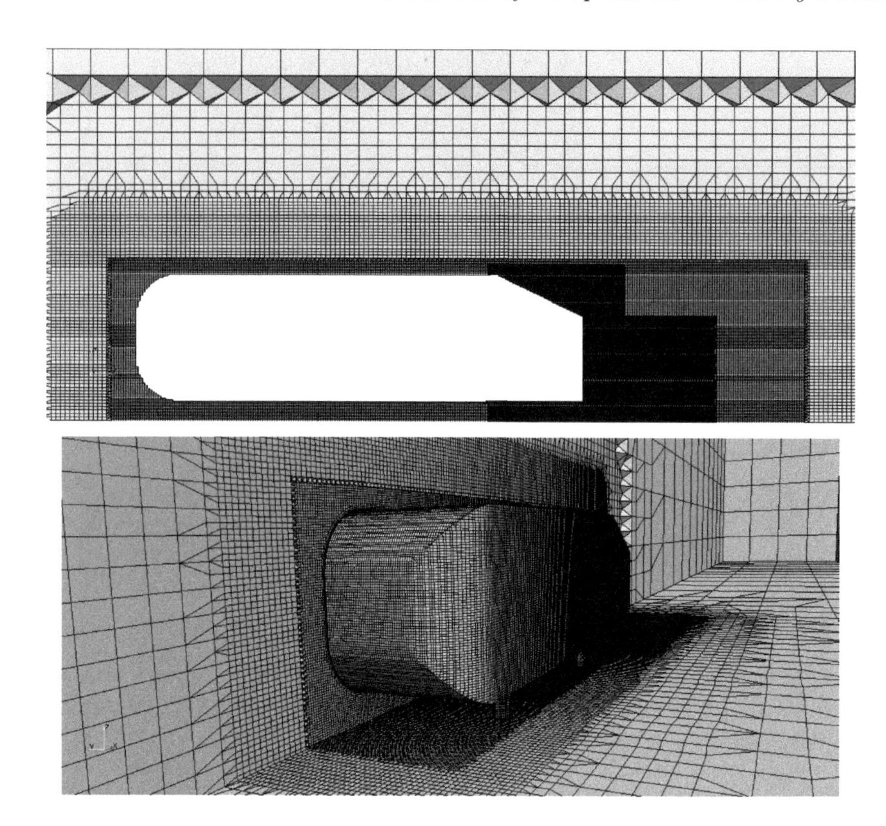

FIGURE 7.18
(a) Side view with refined zones (top), (b) front view (bottom), for Ahmed vehicle geometry, OpenFOAM's snappy hex mesher.

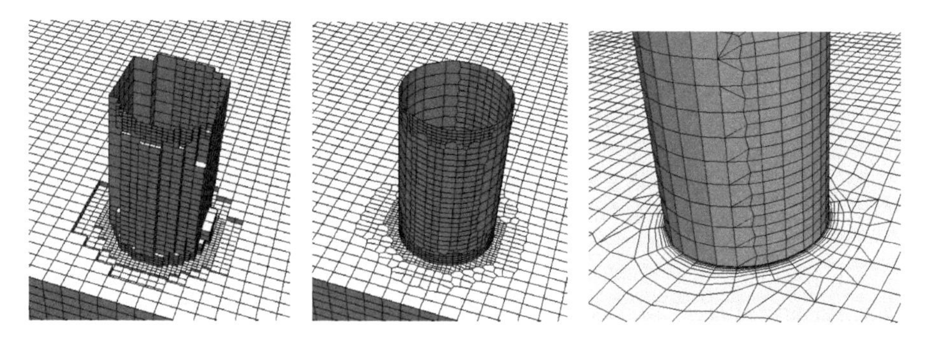

FIGURE 7.19
(a) Castellated mesh around the floor stilt (left), (b) repaired mesh around the floor stilt (middle), (c) mesh with viscous layers around stilt for Ahmed vehicle geometry, OpenFOAM's snappy hex mesher.

rather than tetrahedra saves a large number of elements that would need to be stored and faces where fluxes need to be computed, which more than compensates for the poor efficiency of block refinement. A disadvantage is the poor mesh quality at level interfaces and near the geometry, which may have to be compensated with a finer mesh.

7.4.5 Hybrid mesh generation for viscous flow

The most distinguishing feature of viscous flow is the existence of very thin layers near no-slip walls where the velocity changes rapidly from freestream values to zero (cf. Sec. 6). These layers exhibit a very strong gradient normal to the wall, but much less change in the directions tangent to the wall. An appropriately scaled mesh cell, hence, has a height normal to the wall that is much smaller than the dimensions in the tangent plane.

The required distance from the wall of the first point is discussed in Sec. 6.4.2. It is governed by the chosen approach of near-wall modelling as discussed in Sec. 6.7 as well as the Reynolds number of the flow. The tangential dimensions are governed by the streamwise changes in the flow, which are governed mainly by the geometry and can be estimated considering the flow to be inviscid.

This anisotropy in the cell dimensions is called the *aspect ratio* of the cell. For hexahedral cells it is conveniently measured as the ratio of the longest edge compared to its shortest, a measure that works well if the cell faces are nearly orthogonal to each other. For tetrahedral cells a better measure can be to compare the volume of the cell to the volume of the equilateral triangle with the maximal edge length of the cell.

The mesh generation algorithms for triangular and tetrahedral isotropic meshes presented in Secs. 7.4.1 and 7.4.2 are mature, robust and mostly automated, but not very suitable to produce high-aspect ratio cells as needed to capture layers. Similarly, the unstructured hexahedral methods of Sec. 7.4.4 fix the aspect ratio and direction of the cells with the background grid.

The most popular approach to include stretched boundary layers in the grid is to use a quasi-structured grid near the wall and cut that grid out of the background grid in a very similar way as was done for the geometry. Fig. 7.20 shows an example.

In the case of tetrahedral grids, the surface is tesselated with triangular faces. The typical approach here is to artificially inflate the thickness of the body by inserting a layer of prismatic cells between the triangular boundary faces on the wall and the matching triangular faces of the tetrahedra. Fig. 7.21 shows an example of viscous layers added on the channel walls and flap in a mesh for a damper created with the open-source mesh generator enGrid.[3]

Similarly, in the case of hexahedral unstructured grids, the quadrilateral boundary faces can be inflated to produce stacks of flat hexahedra as shown in

[3]http://www.engits.eu/engrid.

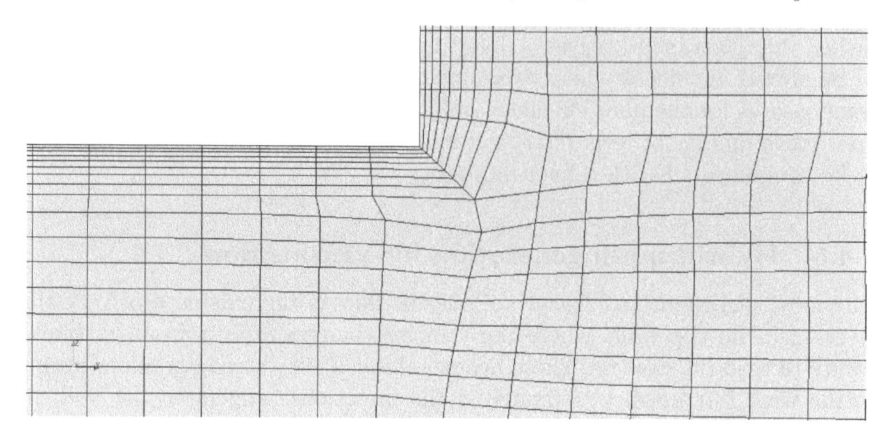

FIGURE 7.20
Stretched boundary layer mesh for RANS computation, Ahmed vehicle geometry, OpenFOAM's snappy hex mesher.

Figs. 7.19 (c) and 7.20. In both cases it is important that the user correctly sets the height of the first cell off the wall and defines the correct stretching such that the appropriate number of cells is within the boundary layer as discussed in Secs. 6.7.1.1 and 6.7.1.2. The size of the anisotropic cells at the outer edge of the viscous layer should match the size of the cells of the background grid to avoid jumps in cell size.

Additionally the user needs to check the quality of the interface between the viscous layer mesh and internal mesh. Problems similar to issues with O-meshing around sharp corners as discussed in Sec. 9.1.5 may occur at corners, regions of strong curvature, intersection of viscous layers with each other where walls are in proximity or where the geometry is very oblique to the Cartesian grid lines.

7.5 Mesh adaptation

The preceding sections on mesh generation have laid out how important it is to tailor the mesh to the features of the flowfield that need to be captured. It is now also apparent how cumbersome it is to control the local mesh size, aspect ratio and its orientation. Typically mesh generation, hence, proceeds in a number of iterations where the user re-generates an improved and refined mesh after inspecting a solution on a previous mesh design.

This process can also be automated, to some extent. An algorithm can estimate the discretisation error in a preliminary solution and predict local mesh scales for an improved mesh. The problem lies mainly with the error

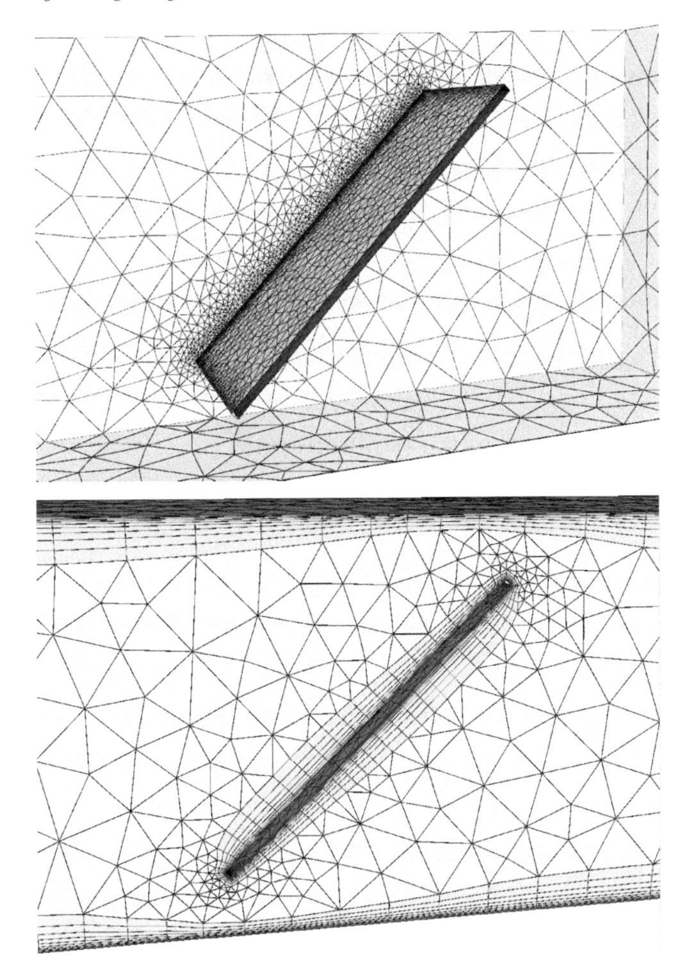

FIGURE 7.21
Tetrahedral mesh for viscous flow around a damper plate produced with en-Grid. Top: initial isotropic tetrahedral mesh. Bottom: mesh with inserted anisotropic prism layers in the viscous layers. (Images courtesy of engits.eu.)

estimation: simple error sensors based e.g., on first or second gradients of the solution fail to capture the non-linear coupling between the equations and between different regions in the flow. This approach is e.g., unaware how the error produced at the leading edge of an aerofoil is transported downstream along the profile by the flow as shown in Sec. 9.1.

There are methods such as the adjoint approach in goal-based mesh adaptation that take the coupling in the flowfield into account, not discussed here; however, they work with linearised models and hence are valid only for infinitesimal perturbations. They cannot capture the non-linear effects either.

The lack of robust error estimation is one of the main problems in CFD, as further discussed in Ch. 8. Hence, while the automated mesh adaptation procedures may help, they still need to be managed by a user who complements the analysis with her/his knowledge about fluid dynamics.

Three principal approaches to mesh adaptation can be distinguished, mesh movement or r-refinement discussed in Sec. 7.5.1, local mesh refinement or h-refinement discussed in Sec. 7.5.2 and finally a local adaptation of the order of accuracy p of the discretisation or p-refinement. The latter is typically not useful for finite volume schemes that are limited to first- or second-order accuracy, but is often used in specialised higher-order discretisation methods for fluids. This approach is not discussed here.

7.5.1 Mesh movement: r-refinement

In r-refinement vertices are relocated; the adaptation algorithm changes the location vector r of each vertex. The topology of the mesh is left unchanged, i.e., the elements and nodes remain connected as they are. The connectivity table as shown in Fig. 7.12 and Table 7.1 remains as is. This keeps the number of vertices and cells identical, and hence the overall computational cost, but it contracts cells into areas of interest while reducing mesh density in areas of lower importance.

Fig. 7.22 shows an example of r-refinement of a structured grid to improve the resolution of a pollutant plume that is advected from a source at the centre in a north-westerly direction. We can observe that the mesh

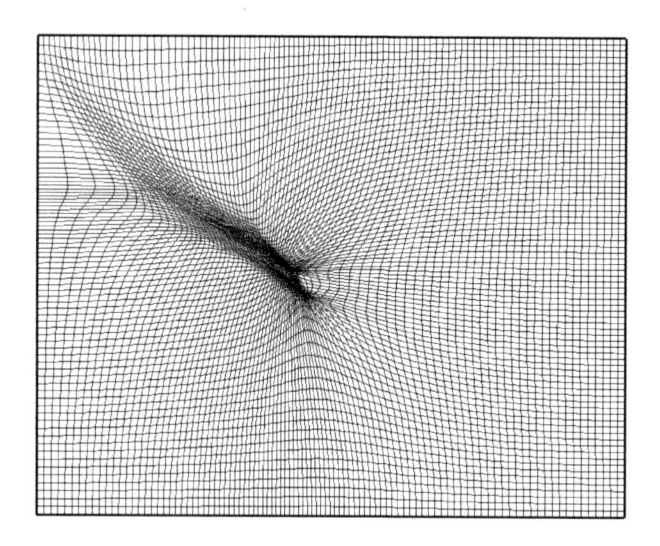

FIGURE 7.22
Example of r-refinement to capture a pollutant plume. (From Garcia-Menendez et al., Atmosphere, 2011.)

deformation adversely affects mesh quality; typically, only limited mesh movement is possible.

A very useful application of r-refinement which most CFD solvers offer is the adjustment of the wall-normal spacing for turbulent boundary layers in order to achieve an appropriate distribution of y^+ (cf. 6.7.1).

7.5.2 Mesh refinement: h-refinement

For more significant changes in the local mesh density we cannot keep the mesh topology fixed but need to insert or remove elements. One of the major advantages of unstructured meshes is that the mesh can be modified locally without affecting the whole domain. In structured meshes the refinement of a single cell into four or eight smaller ones leads to two additional grid lines which have to be continued through the entire domain. In an unstructured quadrilateral or hexahedral mesh the refinement of a cell into four or eight children will only affect its immediate neighbours, but will create cells with hanging nodes. Examples of this can be seen in Fig. 7.18. If the solver can handle the unusual polyhedral elements they can be left in the mesh, or the polyhedral cell needs to be broken up into simpler types.

Adaptive grid methods exploit this advantage. Given an initial coarse mesh an initial solution is computed. Owing to the discussion in Section 4.3.1, the magnitude of the artificial viscosity can be estimated: it is large where the jumps between cells are large. A good indicator or *adaptation sensor* is therefore to refine the mesh where the product of mesh size and flow gradient is large, i.e., the cell jumps are large.

It is very difficult to define an absolute threshold on the sensor value that should lead to refinement of a cell. In order to limit the number of cells in the simulation usually a percentage is fixed. The cells of the original mesh are ranked for their sensor value and the top percentage is refined. A number of adaptation steps can be performed. The limiting factors are usually the available computer performance and storage.

An example of a refined mesh is shown in Fig. 7.23. It shows an unstructured triangular mesh around a NACA 0012 aerofoil. The flow is inviscid at Ma 0.85 at an angle of attack of 1°. The initial grid with around 2000 triangles is shown on the top left. The bottom left shows a very fine solution, the original grid from the top left refined everywhere 5 times to around 1 million cells. It exhibits a strong shock on the upper surface at around 3/4 chord and a weaker shock on the lower surface around 1/3 chord. The stronger shock produces a stronger loss in total pressure, hence has a lower velocity after the shock than the weaker shock. This leads to a slip line with a discontinuous velocity profile in the wake.

The mesh on the right has been adapted nine times. Each step refined about 30% of the elements based on pressure differences, leading to a total of around 20000 cells. The adaptation has properly identified some regions in need of refinement: the leading and trailing edges as well as the upper

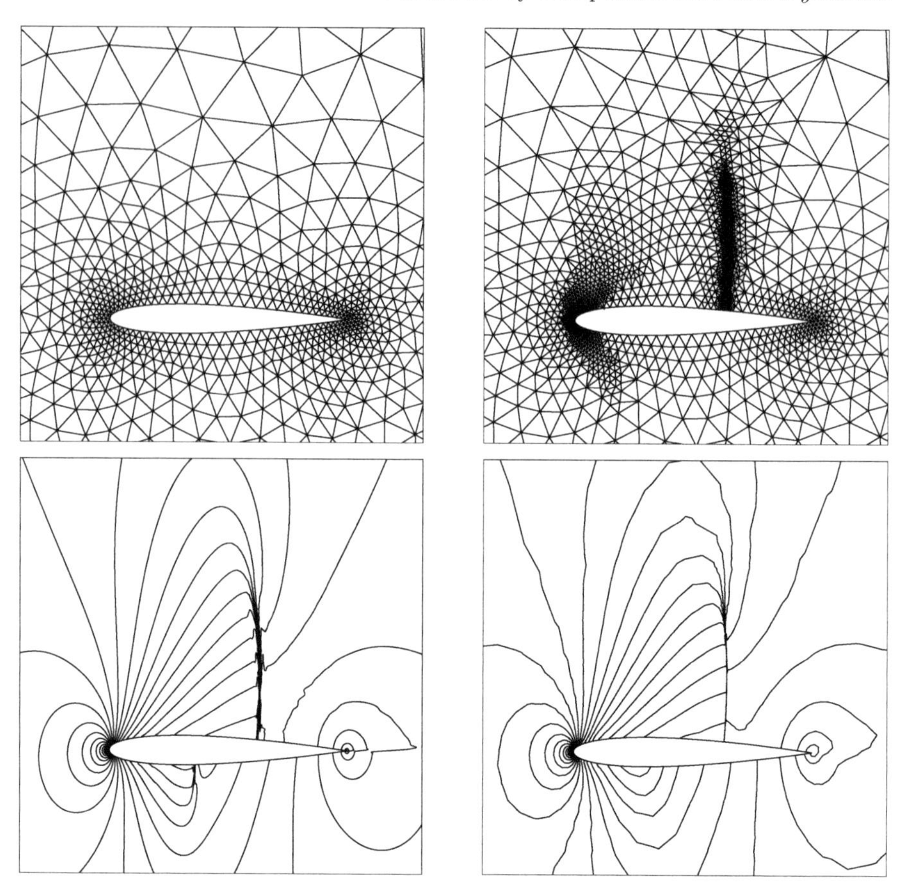

FIGURE 7.23
Adaptive mesh solution. Top left: initial mesh; bottom left: Mach contours of "exact" solution corresponding to five full refinements of the initial mesh. Top right: 9 levels of refinement based on pressure; bottom right: the Mach contours of the corresponding solution.

surface shocks. The corresponding solution is extremely well resolved, exhibiting smooth contour lines and sharp shocks, at around 1/20th of the cost of the fully refined solution.

However, the adaptive sequence failed to capture the shock on the lower surface and also fails to resolve the slip line in the wake. Consequently the predicted lift coefficient is incorrect. The sensor value for the weak shock is too low to be marked for refinement.

What happens in detail in this case: as the mesh width is halved at the shock location, the shock becomes thinner and the gradient through the shock doubles. Each cell is divided into four children; the number of cells along the shock has doubled. This means that as the shock becomes refined, the cells in the shock will become more numerous and sensor values based on jumps will actually remain unchanged after the refinement — the shock is still of the same size and still captured in 2-3 cells. After 2-3 levels of refinement all the available resources for refinement will be absorbed by the shocks. The solution will no longer improve. Note that this process will not terminate in the Euler model — the equations do not take account of the rarefied gas effects that take place in that small area. In the Euler and Navier-Stokes model the shocks are infinitely thin. One has to solve the Boltzmann equations to obtain the correct effects with a shock thickness of a few mean path lengths.

The over-resolution of the shocks in return means that not enough refinement can be applied to the leading and trailing edges. This can mean that the flow-field ahead of the shock is not correctly predicted and the shock location can be incorrect. It is the major pitfall of adaptive methods: very well resolved solutions with crisp features do not mean that these solutions are accurate!

Conversely, there are also features with very strong sensor values which are not relevant to the solution. Compared to the shock on the lower surface, the shock on the upper surface is much stronger, leading to a larger loss in total pressure, a larger rise in entropy and a lower velocity after the shock. Behind the trailing edge these flows merge and exhibit a discontinuity in velocity across the wake. The flow is inviscid; hence, this discontinuity in the exact solution is not diffused and the sharp profile is maintained downstream. The upwind method employed here is able to represent this slipline exactly, provided it runs exactly along the edges of the mesh. However, the gradient-based sensor based on velocity differences would detect a large jump in tangential velocity and trigger refinement. (This has not occurred here because no resources were available due to the shocks.) While in the immediate vicinity to the trailing edge the slip-line influences the flow around the aerofoil and e.g., lift and drag coefficients. Its resolution is not relevant further downstream. The sensor is not aware of how information flows in the problem and will indicate refinement where it does not increase the accuracy of the parameters of interest.

This exposes the major problem with mesh adaptation: there is yet no rigorous error analysis available for the Navier-Stokes equations. The gradient-based sensor used is not a very good indicator of errors in the final solution. When using solution-based flow sensors, features can exhibit a very weak sensor value but still be very important for the flow. Conversely, features can exhibit very strong sensor values and no influence on the features of interest in the flow.

7.6 Exercises

7.1 Discuss the difference between structured and unstructured grids.

7.2 Explain the most important aspects of mesh quality.

7.3 Explain in which situations high-aspect ratio elements are appropriate.

7.4 What particular difficulties do unstructured mesh methods have when dealing with high-Reynolds number flow? Discuss how this difficulty is typically addressed.

8

Analysis of the results

8.1 Types of errors

Erroneous usage by the user is, of course, an error that occurs frequently with inexperienced users. CFD software is complex to use, the discretisation, is challenging and the behaviour of the highly non-linear equations is complex. Confidence in correct use comes with training and practice. To some extent the automated input checking of the CFD code you use may also eliminate certain errors. If in doubt: read the manual.

In many instances the error is not due to wrong or poor practice, but the error is inherent and cannot be eliminated that easily, or may only be reduced as further discussed in the following sections. It is, however, very important that all possible avenues to reduce error are used, within the modelling capability of the solver, or within the memory and CPU limits of the computer. The efforts by the user to reduce the errors should also lead to a deeper understanding how much the remaining errors in the solution affect the quantities of interest that are computed from the flow solution.

Unlike in finite element methods for linear elasticity in structures, there is no rigorous mathematical error-analysis for the non-linear Navier-Stokes equations. The CFD code can't tell you how large the errors in the solution are. The user needs to develop an understanding of the main sources of error, under which conditions they arise, how they scale and how they can be reduced most effectively.

8.1.1 Incorrect choice of boundary conditions

The appropriate application of boundary conditions (BC) is discussed in Ch. 5. Incorrect application of BC can lead to lack of convergence of the solution if contradictory information is specified, or to erroneous solutions if relevant quantities, such as e.g., pressure, is not specified at all.

8.1.1.1 Type of condition and size of the computational domain

Even if the combination of boundary conditions is set up correctly, the choice of boundary conditions can have a significant effect on quantities of interest.

The case study on bifurcation flow in Sec. 9.2 demonstrates this. Boundary conditions are assigned following the example presented in Sec. 5.1.2, Fig. 5.3, with one outlet being assigned a uniform pressure, while the other one is assigned a uniform velocity.

In principle, we have the choice of swapping the two outflow conditions between the two outlets; both choices can satisfy continuity and momentum equations. However, the choice will have an influence on the flowfield near the outlets. Consider the case of placing the uniform velocity condition (with the velocity imposed as flowing out) at the small top outlet; see Fig. 9.11. This uniform velocity profile is actually not correct. It is more likely that the velocity profile has a shape somewhat between the one in the neck shown in Fig. 9.15 and a fully developed parabolic profile. But we don't know what the profile is, so need to impose something and accept that near that outlet our solution will be inaccurate, corrupted by the poor approximation to the actual profile. The question remains how far upstream toward the neck will this error affect the solution; in particular will the profiles in the neck be affected, which is what we are interested in?

As an alternative, we could swap the two conditions around and impose uniform pressure $p_{out} = 0\,\mathrm{Pa}$ at the top outlet. This condition is not correct either: uniform pressure imposes that there are no lateral forces on the fluid. The momentum in the x-direction across the vessel axis cannot change. In the steady state the only solution continuity then allows is that all fluid particles move in the axial direction. The velocity profile cannot change, which implies a fully developed profile. However as the results show, the velocity profile is not fully developed in this case. This is discussed in detail in Sec. 9.2.5.

Fig. 9.17 compares the velocity profiles in the neck. We can see that the choice of boundary condition does affect the velocity profile in the neck: while the profile for either setup is qualitatively the same, quantitative differences can be observed. In the case study the only fact of interest was whether or not a separation existed. This finding was not affected by the choice of boundary condition. If quantitative information was to be extracted from the velocity profiles, the length of the small, vertical vessel should be increased if possible, either by including a longer length of the actual geometry, or by adding an artificial length of vessel to distance the boundary condition from the area of interest.

8.1.1.2 Lack of boundary information

Another source of inaccuracy can be a lack of knowledge of a quantity that needs imposing.

A typical example in external flow simulations using RANS turbulence modelling is how to impose the value of the turbulent quantities at the inflow boundaries, e.g., the $k-\varepsilon$ model solves two additional quantities for the averaged turbulent kinetic energy k and the dissipation rate ε.

Definite values to specify the level of incoming turbulence are typically not available. In turn quantities such as the intensity of the turbulent fluctuations, length scales or viscosity ratios are estimated. The inflow quantities for the turbulent transport equations are then derived from the prescribed sales by dimensional analysis.

In many cases the solution will be relatively insensitive to the choice of parameters, but as the case of viscous flow over an aerofoil in Sec. 9.3 shows, at high angle of attack the prediction of the flow is very sensitive to the freestream turbulence.

Fig. 9.21 shows the effect of varying incoming turbulence on lift and drag. The flow should separate at an angle of attack of around 16° Specifying a high value of turbulence using viscosity ratios results in the k–ε model failing completely to predict any flow separation. Specifying the turbulent length scale based on a fraction of the boundary layer thickness results in two turbulence models predicting the separation too early.

It may be argued that in the case of the high angle of attack with stall, the turbulence models are actually used outside their range of validity; hence, the sensitivity to inflow parameters could be seen as a modelling error (cf. Sec. 8.1.4). Using RANS models for flows with areas of separated flow is, however, typical industrial practice, as no other approaches are available to simulate turbulence economically. Hence the user has to be aware of model sensitivities to boundary conditions.

8.1.2 Insufficient convergence

Some examples how convergence can be affected by incorrectly set boundary conditions, or by complex and poorly stable boundary conditions, have been given in Sec. 8.1.1. In these cases your flow solver may either not converge at all, or may even diverge.

Problems with convergence may also arise if boundary conditions are set correctly. Incompressible flows and the continuity constraint have been discussed in Sec. 5.2.7.2, and the problem this poses for the discretisation in Sec. 3.6.6. A fully incompressible fluid has "no give". The tiniest perturbation in the velocity at the inlet has to propagate through to the outlet instantaneously to satisfy continuity — whether in the steady state or not. The incompressibility condition very strongly couples the velocities in all grid points to each other. This is also referred to as the *stiffness* of the equations.

This constraint manifests itself in practice through poor convergence of the momentum and continuity residuals. For complex industrial cases the solution residuals can often only be reduced by three orders of magnitude from the residual of the initial solution.

In industrial practice when running large cases with complex flow features, it will be impossible to fully converge the solution to round-off error. The user will have to carefully assess whether or not the achieved temporal convergence is good enough, or — if you prefer — the remaining unsteadiness in the flow field is small enough to no longer affect the quantities of interest such as forces, or flow rates. The following points should be considered.

- To obtain an idea how large the effect on the solution could be the magnitude of the residuals should be examined in absolute values and not normalised by e.g., the magnitude of the residual of the initial solution — which is chosen arbitrarily. If the initial solution is already very good, we may have initialised the flowfield of a second-order accurate model from a first-order accurate simulation. The drop in residuals will be much smaller as compared to initialising with freestream or zero flow.

- The flowfield often is converged in most of the domain, but may exhibit significant changes in some specific areas, which in turn may or may not have a strong effect on functions of interest such as lift or drag. When plotting the 'residual', we actually are looking not at values of the flux balance in particular cells, but what is typically shown is the L_2 norm or root mean square value of all the cell residuals. Strong variations in a small part of the domain may then be not visible in this overall measure.

  A very useful approach is to examine the cell residuals in the same way in which we examine the velocities or pressures in the flowfield. Examples of post-processing are given with the case studies in Sec. 9. The analysis of the spatial variation of the residuals will highlight if there are areas in the domain which have not converged. The user can then decide whether this can be tolerated, or whether it needs to be remedied.

- Instability can be caused by poor mesh quality or insufficient mesh density. Investigating the magnitude of the local residuals can point out the areas where the flow solver has difficulties to converge and the mesh quality and density in those areas can be investigated.

  On the other hand on a very, very fine mesh for realistic, complex geometries, flow solvers may be able to resolve physical instabilities such as vortex shedding at rounded trailing edges of turbine blades. A mesh that is too fine is typically not a problem encountered in tutorial-level cases.

- Residual convergence is not an objective in its own right. It just helps to indicate whether or not we are close to the steady state. Of interest are some engineering objective functions that are derived in the post-processing from the solution, such as lift, drag, mass flows, reaction rates, etc.

  Monitoring the convergence of these objective functions is hence more relevant than considering the residuals. The iterations can be halted if an objective of interest evaluated at each iteration no longer changes to the level of expected accuracy. Expected accuracy could e.g., be an engineering accuracy of 3 digits.

  Similarly, we may consider the convergence in time of the solution at specific monitor points that we can place at locations possibly affected by instability, such as the wake behind a bluff body. The fluctuations of the pressures and velocities in these points can provide quantitative information about the level of unsteadiness that remains in the flowfield.

- In most practical cases with incompressible flow, and to a lesser extent with compressible flows (which possess compressibility and are hence less stiff), the flow solver will only converge to *limit cycles*, more or less regular oscillations of residual around some average, constant value. Running further iterations advances the flowfield to a different state in that cycle, but will not make it steady.

  In such a case the user will have to monitor the value of the objective functions over the cyclic variation of the residuals. If the objective does not vary, one can consider the case converged in time. If the objective does vary, the user may consider to average the value of the objective of the cycles. However, in this case we are actually no longer dealing with a steady-state solution but an unsteady flow. It must then be considered whether there is a relevant physical phenomenon behind the oscillations and whether that should be resolved with accuracy in time in an unsteady computation.

8.1.3 Artificial viscosity

As described in sections 4.3.1, 4.3.2 and 4.5.3, the numerical discretisation used in CFD requires the addition of a non-physical artificial viscosity (a.v.) to the mathematical model in order to maintain stability. This viscosity scales with h for a first-order and with h^2 for a second-order method. It can also be interpreted as being approximately proportional to the jump between cells in a finite volume method. Artificial viscosity will also be present in regions of poor grid quality, i.e., large jumps in cell sizes, obtuse cell angles or mesh irregularity (see Sec. 7.1).

When generating the mesh before running the simulation an experienced user will be able to identify the regions where the conditions are 'ripe' for large values of a.v. These are

- regions with strong gradients in the flow, e.g., stagnation points,
- regions with moderate gradients and inappropriate mesh resolution,
- regions with poor mesh quality.

The very first step that can be taken to reduce artificial viscosity is to use a numerical discretisation with the highest order of accuracy available in the solver. Starting with a first-order accurate simulation is a good idea, as it is inexpensive and stable. This allows to get an idea of the flowfield, identify where relevant features occur which may need specific mesh refinement, allows to check the correct setup of boundary conditions and can serve as an initial solution for a higher-order simulation. All flow solvers will typically offer a second-order accurate discretisation, and the next step should be to switch to that after moderate convergence of the first-order scheme. As pointed out in 4.5.3, the convergence rate of a second-order scheme will be lower. Any issues with stability already present with a first-order scheme will be exacerbated.

Some solvers also offer schemes labelled as "third-order" accurate; however, the schemes only exhibit this accuracy on one-dimensional flows with regular structured grids. In practice, in multi-dimensional flow that traverses grid lines obliquely, i.e., the flow is not perfectly aligned with the grid, and on arbitrary unstructured grids these schemes are not more accurate than second-order.

The user also needs to inspect the solution for signs of excessive artificial viscosity. Artificial viscosity acts very much like a viscosity in that it dissipates or smears strong gradients in the flow. In the case of inviscid solutions such errors can be identified straightforwardly as discussed in Sec. 9.1.

In viscous solutions it is often very difficult to distinguish the effects of physical and artificial viscosity, unless, of course, the mesh is extremely coarse. In this situation the user needs to assess the magnitude of the artificial viscosity by conducting a *mesh refinement study* as discussed in Sec. 8.2.

8.1.4 Modelling errors

Modelling errors are much more difficult to identify as they can't be "refined" away. These errors will persist even on very fine meshes. Arguably the most common modelling errors arise from the widely used RANS turbulence modelling, as discussed in Sec. 6.5.2.

Models in general are only valid for a limited range of parameters, e.g., a turbulence model will require a specific wall-distance y^+ of the first node next to the wall depending on whether it is run with or without wall functions (see Sec. 6.7 for details). The user has to ensure that the model is applied using the correct parameters and within the range of validity. A severe deviation from the required range would then be a user error that can be rectified. Practical mesh generation, however, will make it very difficult to match the wall distance on the entire surface, so some mismatch has to be accepted.

More importantly, though, RANS turbulence models rely on particular assumptions about the flow field, e.g., that the boundary layer is attached, fully developed and has zero or only mild pressure gradients. In stagnation and separation points this is not satisfied, and "all warranties are void". The models often perform acceptably outside their original range of validity, and it is up to the user to perform a validation of the performance of the model in that particular flow situation against known experimental or theoretical results; see Sec. 8.3.

In many cases modelling errors can be reduced by switching to more sophisticated modelling with slightly higher cost, or by selecting a model that is specifically suitable for that particular flow. In other cases more sophisticated modelling will dramatically increase computational cost as e.g., when switching from RANS to LES (Sec. 6.5.3), which may not be affordable.

8.2 Mesh convergence

In a consistent discretisation the truncation errors vanish as the mesh width approaches zero. Ideally the mesh width should be reduced until the truncation errors, which include artificial viscosity, have become small enough to become negligible in absolute value, or at least negligible compared to other irreducible sources of error such as modelling errors, or have become small enough for the desired accuracy. It is not easy to compute the size of artificial viscosity in each cell, and much more difficult to then consider the combined effect of the contributions from all cells to the overall objective functions such as lift or drag.

If the mesh is repeatedly refined, the results can be compared between the fine and the coarse mesh. If the difference is larger than the desired accuracy, further refinement is required. Examples of mesh refinement studies can be seen in the case studies in Sec. 9.1 and Sec. 9.2. Clearly, comparing just two solutions may lead to erroneous conclusions. With at least three levels of refinement trends can be confirmed. Also, mesh refinement needs to be substantial to produce effects. Consider a three-dimensional mesh that is uniformly refined from 1M cells to 2M. While this suggests a factor of two, the mesh width is only reduced by $\sqrt[3]{2} = 1.26$. The artificial viscosity is proportional to the mesh width for first-order accurate schemes, not the volume; hence, this minor refinement may not produce significant changes in the objective function.

For smaller cases and in two dimensions it is very often possible to run a complete mesh refinement study. The state one reaches when the results do not change any longer when the mesh is refined is called *mesh convergence*. For three-dimensional cases it is often not possible to achieve this. The grids simply become too big. It is still very useful, though, to do as many refinement steps as feasible to assess how sensitive the solution is to mesh refinement.

Another aspect to keep in mind is that some models we use have specific mesh size requirements, e.g., when using turbulence with wall functions, the first mesh point must be in the log layer, $y^+ \geq 30$. If a user were to refine the near-wall mesh from $y^+ = 30$ to $y^+ = 15$, (s)he would reduce the truncation error, but dramatically increase the modelling error. Should the use of wall function not be appropriate for this flow, the user would need to switch to a low-Reynolds approach (Sec. 6.7.0.2) and significantly refine the mesh near the wall.

8.2.1 Cost of error reduction

A reduction of errors in a CFD solution can be achieved by improving the modelling, refining the mesh, improving the mesh quality or increasing the order of accuracy of the discretisation. Let us estimate the computational cost of mesh refinement and increased order of accuracy in light of the previous section.

Firstly, consider the stability constraint for the explicit first-order upwind scheme, (3.3), CFL $= \frac{a\Delta t}{h} < 1$. The advection equation transports information with speed a to the right; hence, the CFL condition limits the time-step such that the information can at most cross one cell. Running on a finer grid will require the information to cross the domain in more cells, i.e., more iterations are required to convect an error mode to the outflow boundary where it can exit. As the mesh is refined, not only does the cost of each iteration increase because more cells need to be updated, it will also take more iterations for errors to traverse the domain and be swept out through the exit boundary, hence there are more more iterations to converge the residuals to a given level.

Halving the mesh width in each direction in a 3-D calculation will multiply the number of cells by 8, leading to 8 times the amount of work per iteration. This holds if the work spent on the boundaries and other overheads are negligible compared to the work inside the domain. Another factor of 2 is added by considering that convecting error modes across the domain takes twice as long. Overall one has to expect a factor of 16 in work when refining a 3-D mesh by a factor of 2 in each direction. In practice this factor is found to be 1.5-2 larger.

While implicit time discretisations are not subject to the stringent time step limitations of explicit schemes, the solution of the larger linear systems of equations arising from finer meshes also takes longer, and the equations become 'stiffer'. Multi-grid methods (Sec. 4.5.4) are often used to counteract that, but these methods do not show perfect efficiency on the unstructured meshes that are typically used in CFD. On a finer mesh we will need more iterations to converge.

In summary, when refining the mesh the computational cost increases by two factors. One factor is linear in the number of cells, as each cell requires work and memory. An additional factor that can be substantial is due to either a time-step reduction in explicit or unsteady flows, or a reduction in convergence rate when using implicit schemes.

8.3 Validation

After the basic error analysis in the preceding sections has been exhausted, the user is left with colourful graphs which look authoritative but might still contain large errors.

An important way of assessing the accuracy is to run a CFD solution of a very similar case for which experimental data are available, a procedure called *validation*. The comparison of the CFD solution with the experiment will exhibit errors in that particular case. Similar errors can be expected for the predictive simulations for which no experimental data are available. Similarity of flows here can mean at the very basic building block of the flow, e.g., a boundary layer with small separation. More typically, though, in industrial application where high accuracy and high reliability is required, similarity will mean to compare the simulation of, say, a turbine blade under specific flow conditions with a specific type and density of mesh to a slightly different design under the same flow conditions using the same type and density of mesh.

The analysis of the nature of these errors will reveal errors which may be remediable by using better mesh resolution or an improved turbulence model. It will also highlight errors which are unavoidable under the given circumstance, e.g., the necessary mesh resolution is not affordable or the case would take too long to run. At times there may be no affordable turbulence model that could predict the turbulent aspects of the flow with sufficient accuracy.

The remaining errors have to be acknowledged and considered with the solution. It is very important to realise that no CFD solution is without errors, and that the current state of the art does not provide a reliable estimate of the errors in the solution. The user must always exert critical judgement about the accuracy of the solution.

If information is sought in absolute terms, e.g., the drag of a body, the errors can be considerable. In this case very fine meshes need to be run. Solutions need to be very carefully validated against very closely related cases for which experimental data are available.

On the other hand, in engineering design very often only relative information is required. Consider e.g., the case of optimising an inlet for a mass-produced pump where cheap manufacturing results in high tolerances. The modelling error due to the assumption of a perfect, zero-tolerance geometry may be much larger than the discretisation error. What may be more of interest is whether the proposed design change improves the efficiency of the pump or not. Even when large absolute errors are present, relative predictions may be correct and lead to a good design improvement.

8.4 Summary

- All results need to be critically judged.
- User errors can be addressed by carefully checking the flowfield even on coarse meshes.
- Mesh refinement studies are essential to ensure that truncation error is reduced to be acceptably small or reduced as much as can be afforded.

- Modelling errors can not be removed by refinement, but may be affected by changes in the mesh such as wall distance for turbulence models.
- Where possible validation should be performed by carrying out a simulation with the same parameters on a similar flow where reliable data are available.

8.5 Exercises

8.1 What types of errors are typically present in CFD solutions, and what can be done about them?

8.2 Explain the steps a CFD user should undertake to gain confidence in her/his CFD solution.

9

Case studies

The preceding chapters have laid out the essentials of CFD that a beginning user needs to master. In this chapter a number of case studies are presented that investigate specific aspects:

Mesh refinement and order of accuracy: Sec. 9.1 looks at inviscid, incompressible flow over a well-known aerofoil geometry to investigate the effects of mesh refinement and compares results with first- and second-order accurate discretisations. A mesh convergence study is performed and errors are analysed.

Mesh convergence study and boundary conditions: Sec. 9.2 presents the flow through an idealised blood vessel bifurcation with one inlet and two outlets. The effect of boundary conditions is examined, and a mesh convergence study is performed. In this case, the mesh convergence study does achieve convergence on the presented meshes.

Turbulence modelling of boundary layers and its limits: Sec. 9.3 considers standard RANS turbulence modelling with wall functions, as is most widely used in industrial applications. The modelling is applied to the turbulent flow around the NACA0012 aerofoil already studied in invisicd flow in Sec. 9.1. The validity of RANS turbulence modelling is examined for low angles of attack, where the flow remains attached, and at high angles where the flow is expected to separate.

9.1 Aerofoil in 2-D, inviscid flow

9.1.1 Case description

Aerofoil flows are, of course, relevant to aerodynamics of aircraft, but also to race car wings, to bird flight, to flows around turbine blades and e.g., to flows in microturbines used in heart-assist devices. More generally, the flow around a shaped profile exhibits typical flow features such as stagnation points, pressure changes along curved walls and wakes.

The aerofoil is the symmetric NACA0012 profile with a chord length from *leading edge* (most forward tip) to *trailing edge* (most rearward tip) of 1 m. To simplify the flow here we neglect viscosity, i.e., we solve the Euler equations (cf. Sec. 2.3.2); the viscous case is dealt with in Sec. 9.3. The aerofoil is at an angle of 3° against the freestream flow. We could, of course, place the geometry at this angle of attack, but it is simpler to change the incoming flow, i.e., we have the freestream flow arrive at 3° from underneath relative to the symmetric profile. The speed of the incoming flow is taken as $59\,\mathrm{m/s}$.

The simulations are performed with the OpenFOAM flow solver (Version 2.3) using the SIMPLE discretisation (cf. Sec. 3.6.6) for incompressible flow with first- and second-order discretisations for momentum and pressure (continuity).[1]

At the inlet the uniform freestream velocity with angle of attack is imposed; at the outlet a gauge pressure of zero is imposed; at the walls a slip-wall condition is imposed. For all cases iterations are continued until lift and drag values are constant to three significant digits.

9.1.2 Flow physics

The flow over the NACA0012 has been extensively studied, so we have a very good idea what the results should look like. However, we cannot exactly compare the Euler flow solution computed for this case to physical experiments, as the experiments are affected by viscosity. Viscous flow over the NACA0012 is shown in Sec. 9.3 and is compared there to the experiments.

However, by inspecting the character of the Euler equations and the physical effects these equations can represent, we can expect the correct solution to reproduce the following aspects:

Stagnation point: There is a stagnation point at the front of the wing where the flow comes to rest and all kinetic energy is converted into pressure. Hence we should find in this point the velocity to be zero and the static pressure to be equal to the total pressure, i.e., the sum of freestream static and dynamic pressures.[2]

The stagnation point will not be at the *leading edge*, the forward tip of the profile but slightly underneath. This is due to the lower pressure over the top of the aerofoil, which diverts some streamlines over the top surface.

The aerofoil will also have a stagnation point at the trailing edge, but with a more concentrated, localised increase in pressure due to the much smaller change in flow direction.

Suction peak: After the stagnation point the flow accelerates dramatically

[1]The standard implementation in OpenFOAM of the gradient computation at the wall for second-order discretisations produces only first-order accuracy for the inviscid aerofoil case. The cases produced here ran an additional Green-Gauss correction step for the wall nodes.

[2]As typical in flows of low-density gases, changes in potential or hydrostatic energy, $\rho g h$, can be neglected compared to the other terms in the energy equation.

away from the stagnation point. The streamlines near the profile also experience very strong turning which needs to be brought about by a strong gradient of pressure: the pressure on the profile needs to be much lower than the pressure in the freestream in order for the streamlines to curve toward the lower pressure.

This results in a continuous drop of pressure along the profile downstream of the stagnation point up to a point of minimum pressure, the *suction peak.* This is not located at the point of maximal curvature (which for this profile would be the leading edge) due to the fact that the centripetal force required to keep particles on a curved streamline is related to the radius of curvature, but also the speed of the particles. Note that the acceleration of the flow in this case is not related to the displacement of the aerofoil. This effect is negligible compared to other effects here.

Boundary layer: One might expect a boundary layer on the profile, but not in this case. We have neglected viscous effects. The Euler equations allow streamlines to slip past each other without resistance. Hence, the flow will not obey a no-slip boundary condition; velocity will be non-zero at the wall.

The effects of streamline curvature will be most pronounced right near the wall. We actually expect the highest velocities right at the wall.

Wake: Since there is no boundary layer with loss of mechanical energy, we'd also expect not to see a wake (a trail of diminished velocity downstream of the trailing edge). In this inviscid flow the Bernoulli equation is valid: the total pressure, i.e., the sum of static and dynamic pressures, is constant along a streamline. And since all streamlines enter with the same pressure and velocity, the total pressure should be uniform throughout.

Farfield: With a positive angle of attack the profile will produce lift. The lower pressure over the top will "pull" the streamlines up over the leading edge, while the trailing edge at an angle will send the flow downward. It is in effect this vortical motion swirling the flow in a clockwise motion if the flow comes from the left, and the related change of momentum which produces the lift. The effect of this vortex decays in magnitude only with $1/r$ where r is the distance from the aerofoil. An inlet boundary condition that imposes uniform flow is, hence, incorrect unless placed very far from the aerofoil. The discussion of the setup of the NACA0012 test cases on the NASA website [10] suggests a distance of at least 400 chord lengths, which would require a large number of cells in a structured grid. We shall make do here with 20 chords.

Lift and Drag: We don't know what the lift for this profile in inviscid flow should be. Analytic approximations for thin profiles (but the NACA0012 is not that thin) approximate the lift coefficient as $c_L = 2\pi\alpha$, where α is the angle of attack in radian; hence, here $c_l \approx 0.33$. Any deviations from that value may mean that the simulation is inaccurate, or that the analytic approximation is not good enough.

However, we do know what the drag should be. D'Alembert's[3] "paradox" elucidates this. If a body is immersed in an inviscid fluid, the retarding pressure force on the upstream side is equal to an accelerating pressure force on the downstream side of the body. The net drag in inviscid, attached flow is zero. It was termed a 'paradox' since D'Alembert could mathematically prove that there was no drag in inviscid flow, but drag was observed in experiments with viscous flows.

The analysis of the flow physics above should be clear to any successful student of basic fluid mechanics. Textbooks on basic fluid mechanics will help to fill any gaps.

9.1.3 Meshes

Two different topologies of structured meshes are used, C-meshes and O-meshes, as shown in Fig. 9.2. In structured meshing we can map the actual mesh and its coordinates η, ξ along either mesh line direction onto a rectangular domain (cf. Sec. 7.3.1). The mapping for either topology is shown in Fig. 9.1 for a rectangular grid with M vertices in the direction tangential to the aerofoil, N vertices normal to it. It can be observed that the O-grid has M vertices on the aerofoil surface D, while for the C-grid only about $M/2$ are on the aerofoil along D. $M/4$ are needed for the lines from trailing edge to exit boundary labelled D and F. Placing twice as many points along the aerofoil surface D cuts the mesh width h in half, which in turn reduces the truncation error.

On the other hand we can observe that the mesh quality at the trailing edge in the C-mesh is quite good with nearly orthogonal cells. The cells at the O-mesh at the trailing edge are highly skewed with very large maximum angles. What is worse, a mesh refinement will place the first interior mesh line even closer to the wake line C/E, resulting in even more pronounced skewness of the cells at the trailing edge: mesh refinement can't help to reduce this error but will make the mesh quality worse. The resulting mesh around the aerofoil is shown in Fig. 9.2, details around the trailing edge in Fig. 9.3.

9.1.4 Simulation results for the C-mesh

9.1.4.1 Velocity field

Velocity contours are shown in Fig. 9.4. The top left image shows the solution of the coarsest 128 x 32 grid with a first-order accurate discretisation. We

[3] Jean-Baptiste le Rond d'Alembert, French mathematician, physicist and philosopher, 1717–1783.

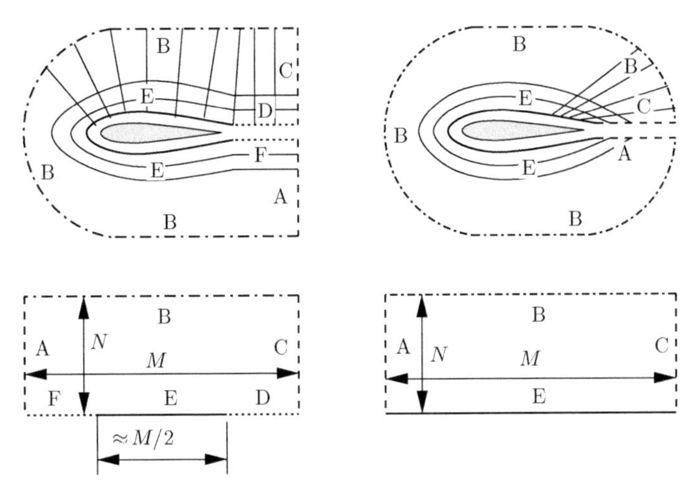

FIGURE 9.1
C-mesh and O-mesh topologies for an $M \times N$ structured mesh. Dash styles and letters indicate how the sections of the perimeter of the grid rectangle are mapped onto the geometry.

can observe a stagnation point and the zone of low pressure over the top of the aerofoil. However, there is a distinct boundary layer, especially along the suction side (top) but also along the pressure side (bottom). These boundary layers merge at the trailing edge and produce a distinct wake. As discussed in Sec. 9.1.2, there should be neither a boundary layer nor a wake in this inviscid flow. We also can observe non-physical oscillations, *wiggles*, in the contour lines which should be smooth.

These errors are produced by the artificial viscosity that is included in the numerical discretisation to maintain stability. In the first-order accurate method used here, it scales with the product of cell size and first gradient (cf. Sec. 4.3.2). It would seem that the largest errors are at the trailing edge, where the 'artificial boundary layer' is the thickest. However, the effect of artificial viscosity is as cumulative as the effect of physical viscosity: the layer grows as artificial viscosity dissipates momentum, and the deficient velocity profile is then advected downstream. The lost kinetic energy cannot be recovered, and as we will see from further analysis, the largest errors are actually found near the leading edge.

The bottom right element in Fig. 9.4 for the second-order accurate solution on the finest 512 x 128 C-mesh reveals much reduced errors. The boundary layer effect is much reduced; the wake is still present, but very diminished. The contours around the stagnation point at the trailing edge have become circular, similar to the contours around the stagnation point at the leading edge.

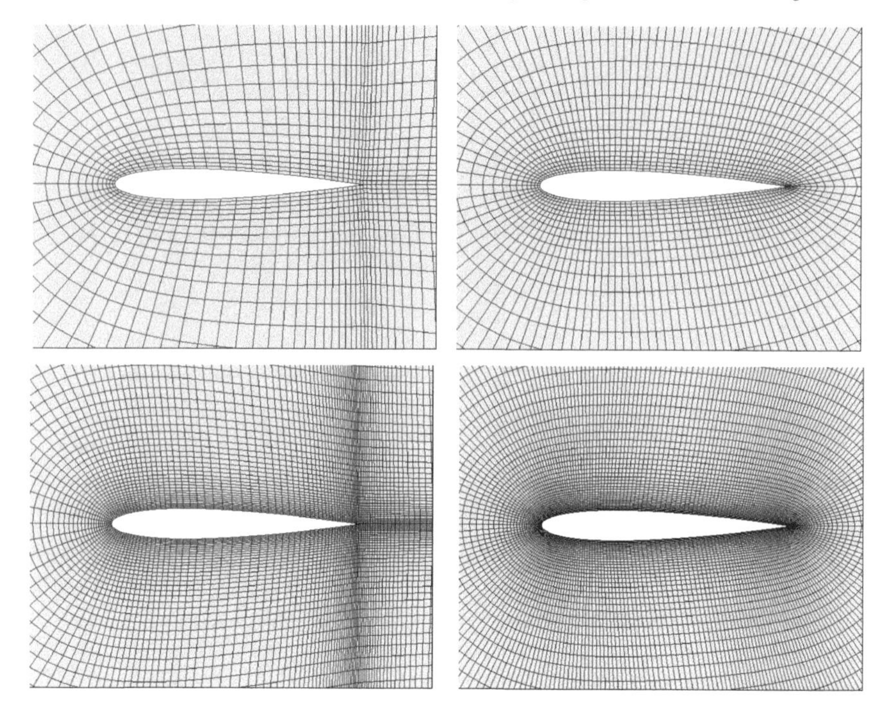

FIGURE 9.2
Closeup of meshes around the NACA0012 aerofoil. Top: 128 x 32, bottom: 256 x 64. The 512 x 128 is too dense to be shown. Left: C-mesh, right: O-mesh.

Refining the mesh (moving down in rows) or increasing order of accuracy (moving to the right column) both show significant improvement in accuracy. It is interesting to observe that the second-order solution on the 256 x 64 mesh (middle row, right column) exhibits a less pronounced boundary layer effect and better trailing edge resolution compared to the first-order solution on the 512 x 128 mesh.

Given this analysis, the second-order accurate solution on the 512 x 128 mesh looks pretty good. Is it good enough, or should we consider an even finer mesh? Let us analyse the solutions in more detail.

9.1.4.2 Static pressure

The loss of kinetic energy, or dynamic pressure, is not only apparent in the velocity contours, but also in the profiles of static pressure in Fig. 9.5. Note that the pressure is drawn with the negative direction pointing upwards. This is customary in aerofoil analysis since the lower pressure over the suction side at the top produces upward lift.

Recall from the discussion of the flow physics of this case, Sec. 9.1.2, that the pressure gradient normal to the streamline, and ultimately the difference

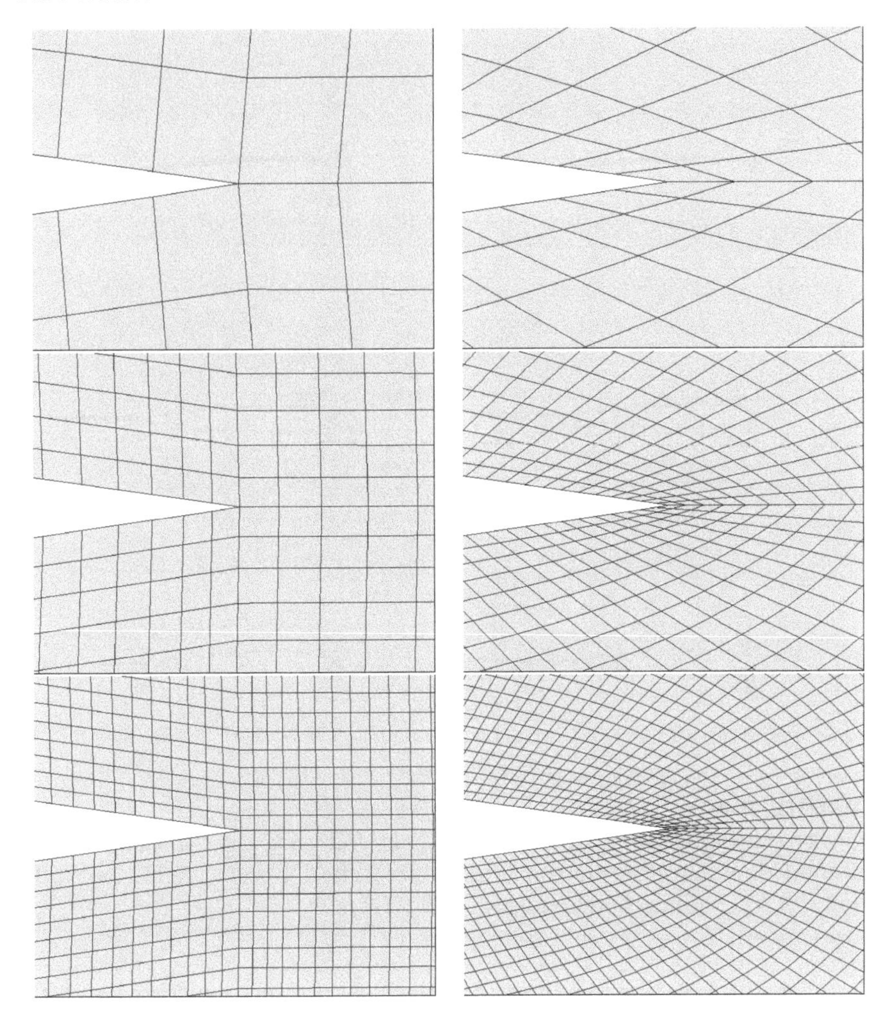

FIGURE 9.3
Meshes around the trailing edge of the NACA0012 aerofoil. Top: 128 x 32, middle: 256 x 64, bottom 512 x 128. Left: C-mesh, right: O-mesh.

between the freestream pressure and the pressure at the wall, is related to the radius of curvature of the streamline and the velocity of the particles travelling on it. If the velocity is reduced, a lower pressure gradient suffices to achieve the same turning of the streamline.

We can precisely observe this in Fig. 9.5. The suction peak (negative static pressure below atmospheric) increases in magnitude with mesh refinement and when switching from first- to second-order accuracy. The more accurate solutions experience lower levels of artificial viscosity, hence a higher velocity near the wall, which in turn results in lower static pressure, or a higher suction

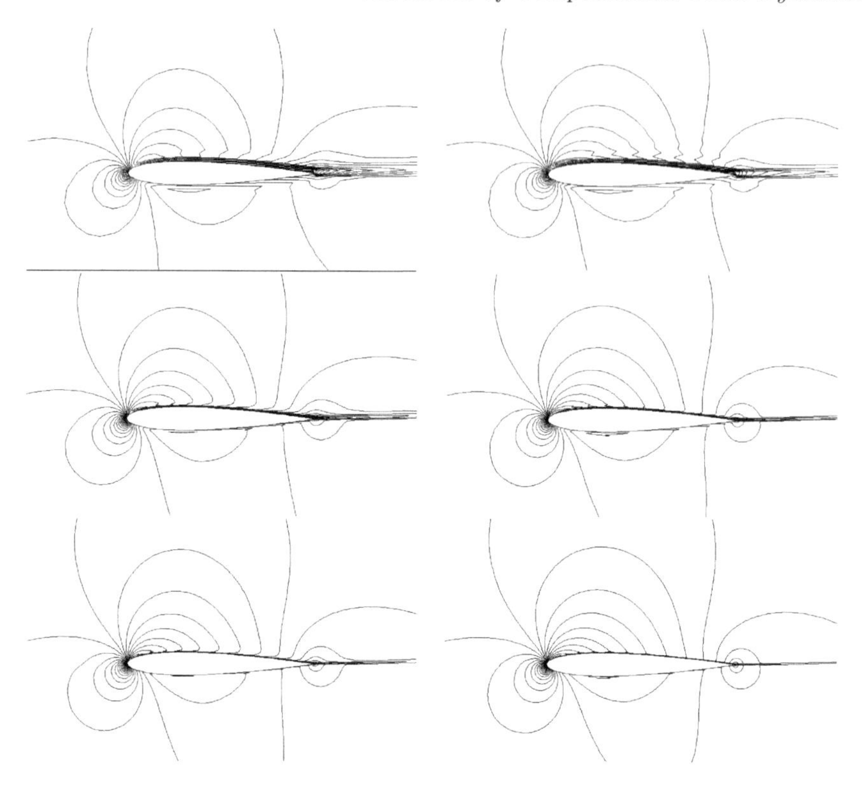

FIGURE 9.4
Velocity contours around the NACA0012 aerofoil using the C-mesh. Top: 128 x 32, middle: 256 x 64, bottom 512 x 128. Left: 1st-order, right: 2nd-order accuracy. 40 contours between 0 m/s and 90 m/s.

peak. The lift force L is the integral of the local pressure multiplied with the component normal to the freestream of the local wall normal $\mathbf{n}_\perp$, $L = \int p d\mathbf{n}_\perp$. The reduction in static pressure that we can observe is hence linked to an increase in lift prediction.

We can observe that refining from 256 x 64 to 512 x 128, the suction peak increases quite a bit, even with second-order accuracy. We will expect that the lift also will change. In Sec. 9.1.6 the lift and drag values will be analysed to determine how large the change in lift and drag still is, and whether or not we can consider the second-order 512x64 solution to be "mesh converged".

9.1.4.3 Total pressure

To quantify the effect of artificial viscosity we can investigate the behaviour of the total pressure p_{tot}.

In this inviscid flow with constant inlet velocity, the total pressure should be $p_{tot} = \frac{1}{2}\rho u_\infty^2 = 2097\,\mathrm{Pa}$ since the freestream pressure at the inlet is the

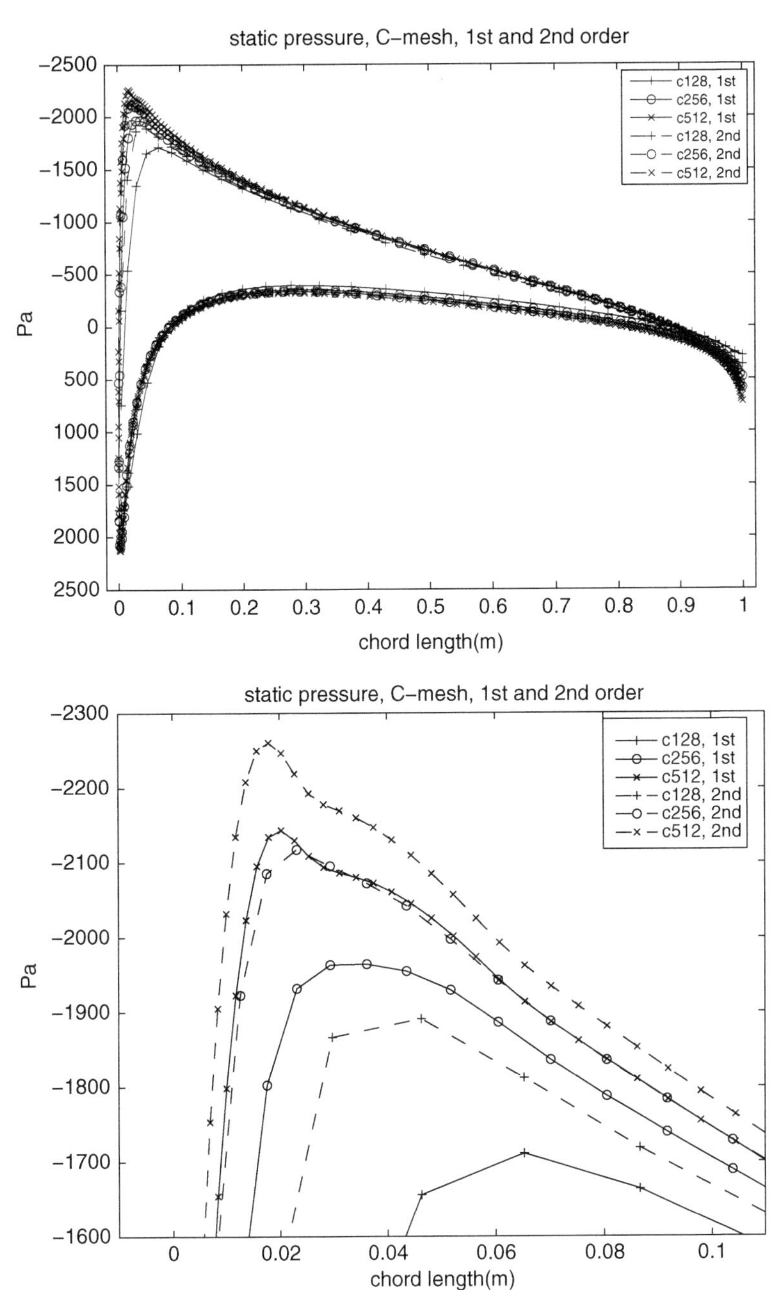

FIGURE 9.5
NACA0012, C-mesh, static pressure over the chord length (top), detail near suction peak (bottom).

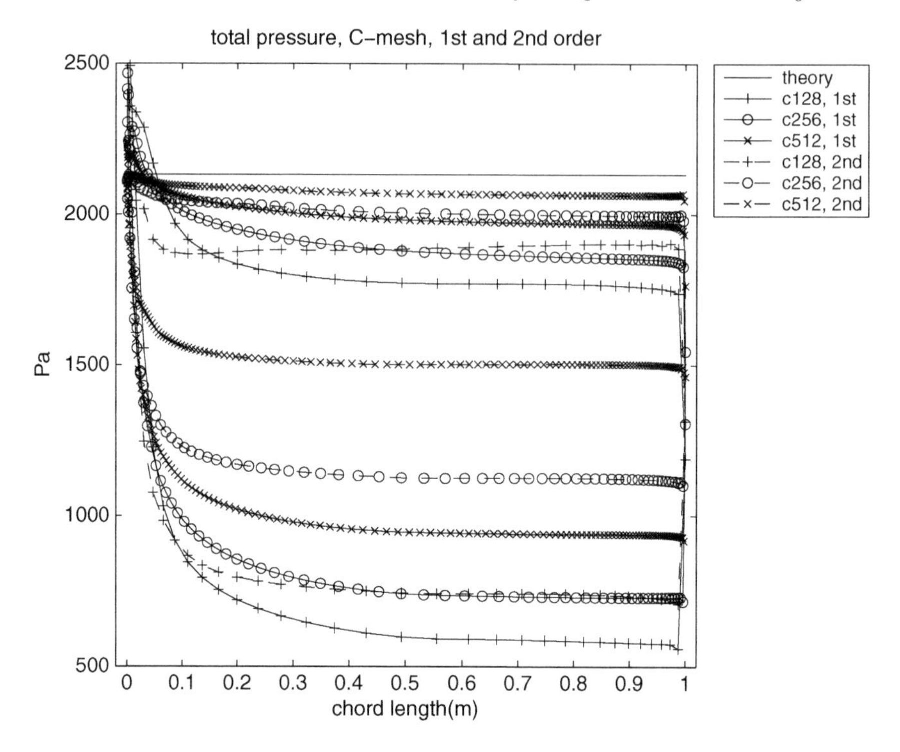

FIGURE 9.6
Total pressure for invisicd NACA0012 on a C-mesh.

same as at the outlet and taken as zero gauge pressure. The actual total pressure along the profile is recorded in Fig. 9.6. The first-order accurate solutions have an unphysical overshoot of total pressure in the stagnation point which is most likely due to the implementation of the solid wall boundary condition in this solver. This is corrected by using the second-order scheme.

Downstream of the stagnation point the total pressure drops off dramatically. At the trailing edge the two streamlines from upper and lower surface join and their total pressure is averaged. As expected, refining the grid or switching from first- to second-order accuracy reduces the total pressure loss.

A much stronger loss of total pressure is observed for the suction side (top of the aerofoil): these are the 6 lower curves. The pressure side also experiences a total pressure loss, but on this side it is much less pronounced. The streamline that travels from the stagnation point over the top, suction side, experiences much stronger acceleration and curvature compared to the streamline travelling over the lower, pressure side. The jumps between the cells that the suction side streamline traverses will be much stronger; hence, there will be higher levels of artificial viscosity. This results in a much higher loss of total pressure.

The loss occurs primarily over the first 25-50% of the chord length. Total pressure remains mainly constant in the rear half of the aerofoil. The acceleration and curvature of the streamlines are most pronounced near the leading edge; hence, it is there that the highest levels of artificial viscosity are applied, and hence it is there that the largest losses of total pressure are observed. Once the kinetic energy is dissipated, it cannot be recovered even if the mesh is fine enough. The total pressure remains at near constant value on the rear half of the aerofoil.

Each of the symbols on the curves corresponds to a mesh point. We can observe that the mesh is clustered (concentrated) toward the leading and trailing edges, as also visible in the mesh plots in Fig. 9.2. The near-constant values of p_{tot} in the rear half indicates that the mesh is fine enough there to resolve the behaviour of p_{tot}.[4] The strong loss of p_{tot} after the leading edge demonstrates that it might be advisable to cluster the mesh even more into the leading edge.

9.1.5 Comparison of C- vs O-mesh

In the previous section we have seen that the second-order accurate solution on the 512 x 128 C-mesh is not mesh converged. Static and total pressure show significant changes with mesh refinement from the 256 x 64 to the 521x128 mesh. Sec. 9.1.3 has demonstrated that the O-mesh topology allows to have all of the 512 nodes in the profile-tangential direction to reside on the aerofoil profile, compared to only about half of them for the C-mesh. With the halved mesh width, we'd expect smaller cell jumps and hence reduced artificial viscosity, and ultimately a more accurate solution.

The mesh width normal to the wall and tangential to the wall at the leading edge is shown in Table 9.1. The finer O-mesh around the leading edge manifests itself also in the lower minimal velocities in the flowfield, compared to the C-mesh with the same number of nodes. The smaller cells in the O-mesh average the solution over a control volume with a smaller size; hence, the solution on the O-mesh resolves the stagnation point with zero velocity a bit more sharply.

Fig. 9.8 compares the static pressure for the second-order accurate C and O-mesh solutions; again the pressure axis is negative in the upward direction. As expected, the suction peak increases (decrease of the lowest pressure) with mesh refinement. Also, as expected, for the 128 x 32 and 256 x 64 meshes the O-mesh produces lower pressures. However, for the 512 meshes the peak value is identical between C and O, and the C-mesh retains a slightly higher suction (lower pressure) downstream of the suction peak. Considered in isolation, we could not decide whether the solution on the C512 or the O512 mesh is more accurate, since we do not possess an analytic solution. But the mesh refinement

[4] The discussion of the trailing edge behaviour in the next section will demonstrate that there are relevant flow features which couple the flowfield globally but whose accurate resolution can't be assessed by simply looking at local gradients in the flowfield.

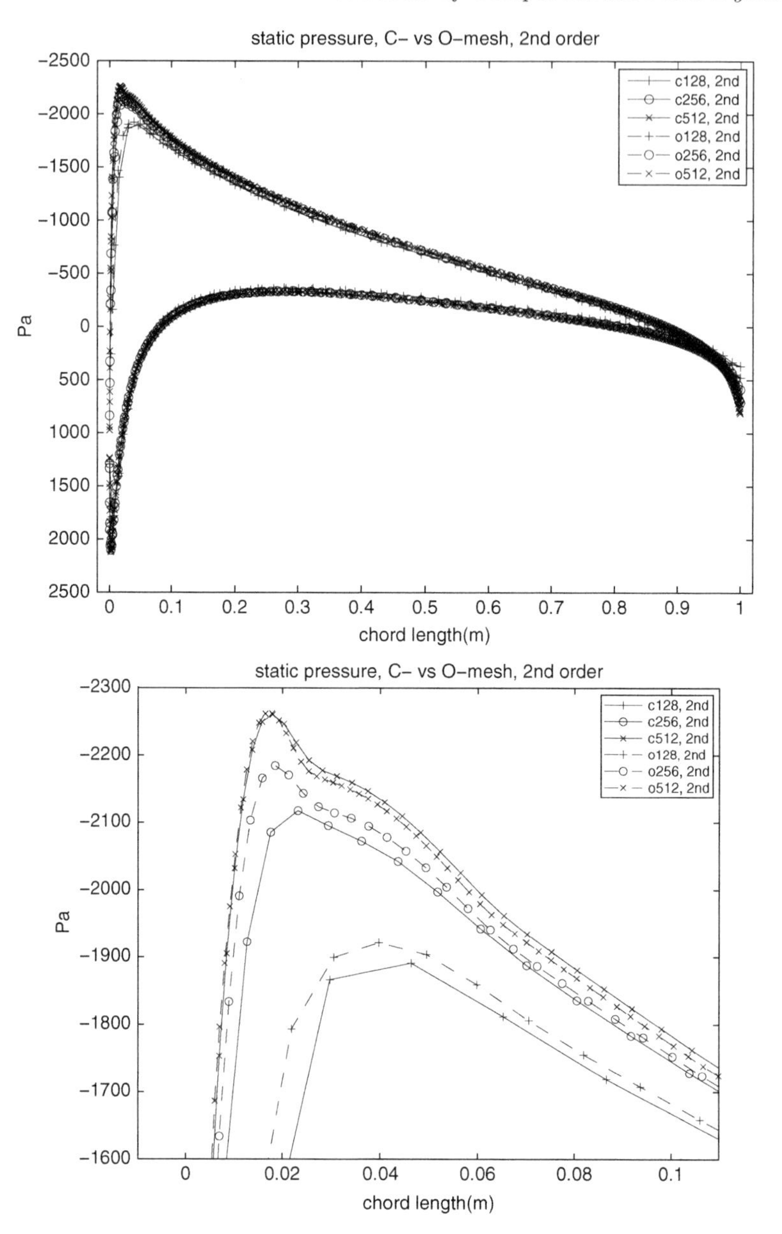

FIGURE 9.7
Static pressure over the chord length (left), detail near suction peak (right). C- vs. O-mesh, second-order accurate solutions.

study shows that the suction peak is enhanced with mesh refinement and that we have not yet reached mesh convergence. Hence, with more refinement a

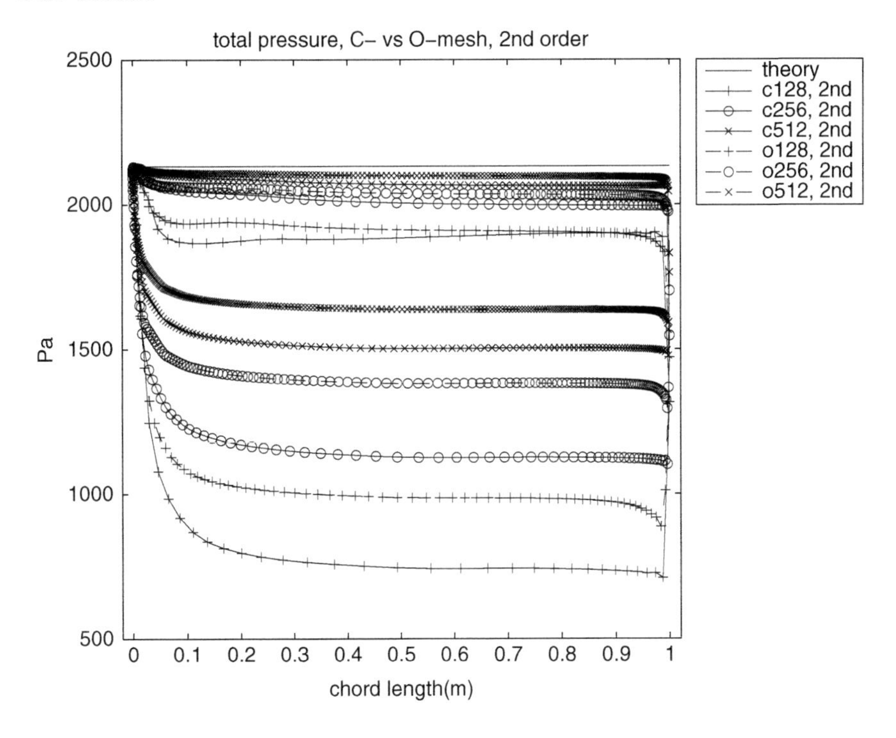

FIGURE 9.8
Total pressure, C- and O-mesh solutions with second-order accuracy.

further reduction in the pressure over the suction side is expected, which leads us to conclude that the lower pressure of the C-mesh is more accurate.

Analysis of the behaviour of total pressure in Fig. 9.8 can shed some light on why this is. In particular let us focus on the profiles of p_{tot} over the suction side shown in the curves with the stronger total pressure loss. The symbols indicate the mesh density. It can clearly be seen that the O-mesh is about twice as fine as the C-mesh along the aerofoil. This manifests itself in reduced losses of total pressure: all O-meshes produce results closer to the theoretical value than their C-mesh counterparts. Both mesh types have a similar, but moderate clustering which leads to a strong drop of p_{tot} over the first 10-20% of chord, and a near-constant value for the remainder of the aerofoil.

Different from the C-mesh solutions, the O-mesh solutions do exhibit a sharp drop in p_{tot} very close to the trailing edge. With mesh refinement this reduction appears more concentrated toward the trailing edge, but the size of the drop remains constant. This error cannot be "refined away". This reduction is due to the poor mesh quality at the trailing edge (cf. 9.1.3): mesh refinement actually makes the skewness of the cells at the trailing edge worse.

One might argue that velocity gradients at the trailing edge are not that strong, compared to the leading edge, and that the poor mesh quality there

T	sz	acc	c_L	c_D	$h_{\|\|}$	$h_\perp$	v_{min}	v_{max}
C	128	1	0.266	0.03000	1.180	1.000	22.4	73.8
O	128	1	0.275	0.02310	0.781	1.000	25.3	75.3
C	256	1	0.305	0.01530	0.226	0.500	15.4	77.8
O	256	1	0.308	0.00977	0.207	0.250	9.3	79.7
C	512	1	0.324	0.00739	0.132	0.250	8.3	80.6
O	512	1	0.323	0.00541	0.113	0.125	4.7	81.4
C	128	2	0.283	0.01270	1.180	1.000	22.7	78.2
O	128	2	0.293	0.01060	0.781	1.000	20.6	78.3
C	256	2	0.320	0.00548	0.226	0.500	12.0	80.7
O	256	2	0.329	0.00355	0.207	0.250	6.3	82.1
C	512	2	0.341	0.00277	0.132	0.250	5.6	82.4
O	512	2	0.338	0.00205	0.113	0.125	3.4	83.0

TABLE 9.1
NACA0012: table of lift and drag coefficients, mesh widths (in % of chord) at the leading edge tangential and perpendicular to the wall, as well as minimal and maximal velocities (m/s) in the flowfield.

should not matter. Unfortunately this argument is too simplistic for the trailing edge of a lifting aerofoil. The entire flowfield is very sensitive to the flow angle at the trailing edge. It determines the loading of the aerofoil. that is the pressure difference between top and bottom. A slightly more downward flow angle at the trailing edge will force the stagnation point slightly further back, resulting in more flow travelling over the suction side, which in turn increases the velocity there which means lower pressure. And this is precisely what we observe comparing the 512 x 128 meshes with C and O topology.

This behaviour is an example of the complex non-linear nature of the flow equations: small local changes can have global ramifications. The user will have to develop an understanding of the behaviour of the flow for his particular geometry, and pay special attention to such critical spots.

9.1.6 Analysis of lift and drag values

Engineers are typically less interested in the velocity or total pressure values. The objective functions that are used to compare two designs are typically integral functions. In this case they are lift and drag as tabulated in Table 9.1.

In this inviscid flow we know that the drag should be zero, as discussed in Sec. 9.1.2. Mesh refinement and higher order of accuracy indeed decrease the

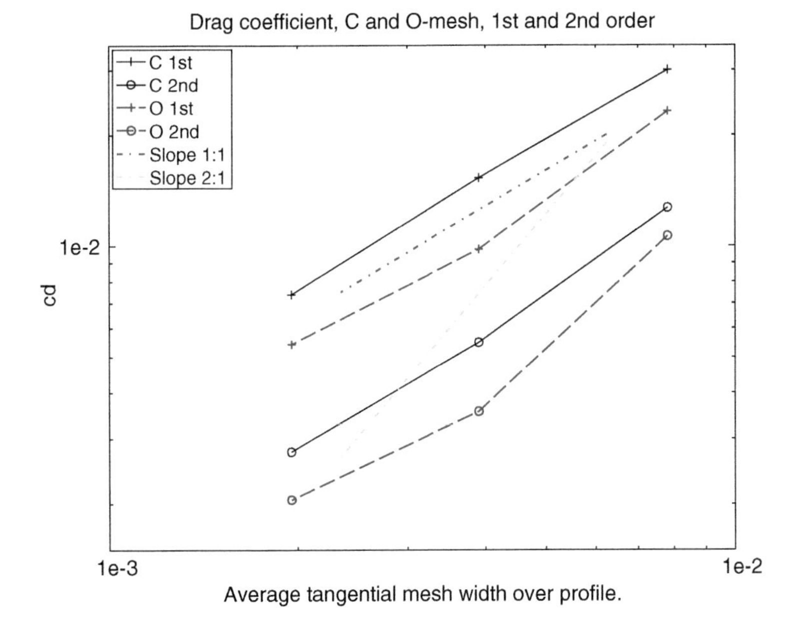

FIGURE 9.9
Reduction in drag coefficient error c_D with relative mesh size.

drag and, hence, the error. The drag of the O-mesh solutions is consistently lower than on the comparable C-mesh which is directly linked to the smaller tangential mesh size of the O-mesh topology. If the objective is to get the drag right, then the O-mesh is the better option.

A second-order accurate discretisation will reduce the truncation error to a quarter when the mesh size is halved. This does not mean that the error in the solution to the system of non-linear equations behaves in the same way, or for that matter the error of an integral quantity such as lift or drag evaluated from that solution. We can measure the order of accuracy of the drag by measuring reduction in drag error in a log-log plot $\log c_D$ vs. $\log h$. The average mesh width h is taken as $1/M$ where M is the number of vertices in the direction tangential to the aerofoil; hence, $1/128$, $1/256$ or $1/512$ for the meshes used. The leading error of p-th order accurate behaviour of the error f of a function can be expressed as $f = ch^p$. A log-log plot shows this as $\log(f) = p\log(h) + c$; hence, the slope of the log-log curve corresponds to the order of accuracy of f.

For this test case the drag error in the second-order accurate solutions is cut in half with every refinement, not to a quarter, as shown in Fig. 9.9. The log-log slope of the drag error for both discretisations is close to one. However, the level of error of the second-order solution is about half as large as the one of the first-order solution on the same mesh.

The analysis of the static pressure profiles in Fig. 9.5 and 9.8. has shown that the lift further increases when refining from the 256 x 64 to the 512 x 128

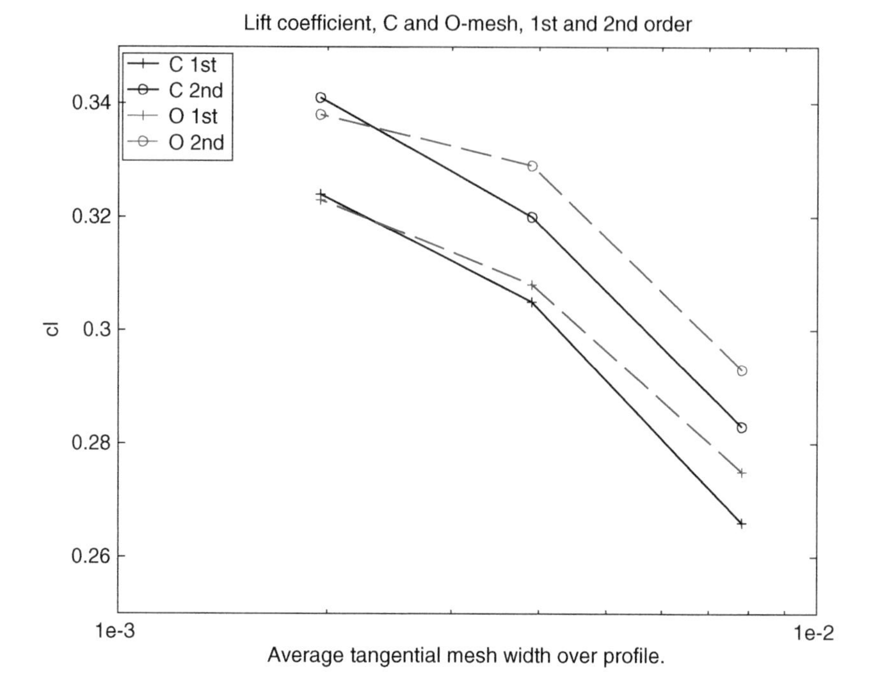

FIGURE 9.10
Change of lift coefficient c_L vs. relative mesh size.

mesh. This is confirmed in the plot of lift vs. log of mesh width (Fig. 9.10). We do not know what the exact lift coefficient c_L should be, and hence cannot determine a lift error. Therefore, the lift axis is linear while the mesh width is in log scale.

The slope of the lift increase is levelling off for the O-mesh when refining from 256 x 64 to 512 x 128, which is due to the poor mesh quality at the trailing edge (cf. Sec. 9.1.5). If an accurate lift value is the most important objective, the C-mesh is the better choice.

With either choice of mesh, the second-order simulation on the 512 x 128 mesh is the most accurate simulation considered here, but at that mesh resolution this case is not yet mesh converged. Further mesh refinement is required.

9.2 Blood vessel bifurcation in 2-D

9.2.1 Geometry and flow parameters

This case considers viscous flow of blood in a simplified blood vessel bifurcation. Flow enters at the left, and exits at the right, and through the small

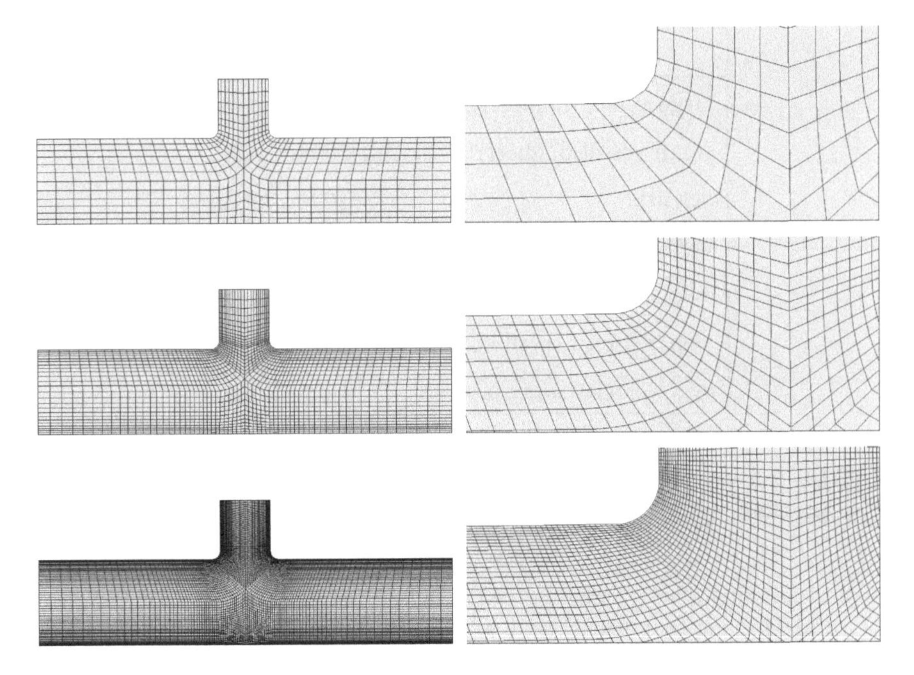

FIGURE 9.11
Three levels of structured meshes for a simplified blood vessel bifurcation. Mesh for the fourth level is not shown. Levels 1 to 3 from top to bottom. Left: complete domain, right: detail around the neck.

vessel at the top, the geometry is symmetric around the y-axis. Fig. 9.11 shows the geometry and the first three mesh refinement levels.

The flow strongly turns from the main horizontal branch into the smaller vessel. There may be a flow separation with a recirculation bubble at the upstream side of the neck. Flow separation and reverse flow are linked to atherosclerosis, a degenerative disease of the vessel wall that may lead to buildup of plaque in that region, with occlusion of the smaller vessel or stroke as possible consequences. Hence in this case the surgeon would like to know from us whether the flow in this particular case will remain attached at the neck, or whether it will separate, in which case he/she may operate to reshape the bifurcation for a better outcome.

To simplify our case, the pulsatile nature of the flow is neglected; the flow rate approximates an average velocity over the pulsatile cycle. The flow is taken as laminar and incompressible with an average inflow velocity of $0.25\,\mathrm{m/s}$ in the main vessel and an average outflow velocity of $0.075\,\mathrm{m/s}$ at the small vessel. As often practised in simulations of blood flow, the shear-thinning behaviour of blood is neglected and a constant viscosity is assumed. The case assumes that measurements of velocity or pressure profiles at the inflow and outflow planes are not available. This assumption is realistic: while

Geometry and flow parameters		
large vessel diameter:	$2 \cdot 10^{-3}$	m
small vessel diameter:	$1.2 \cdot 10^{-3}$	m
density:	1040	kg/m^3
viscosity (Newtonian):	0.04	Pa sec
avg. velocity inlet:	0.25	m/s
avg. velocity small outlet:	0.075	m/s
avg. velocity large outlet:	0.205	m/s

TABLE 9.2
Flow parameters for the bifurcation case.

medical imaging allows us to derive the actual patient-specific geometry from CT or MRI scans (see e.g., Figs. 1.8-1.11), the clinician would only have very limited knowledge of flow rates or pressures. In this case we have, hence, to make assumptions about inflow and outflow conditions. The flow parameters are summarised in Table 9.2.

The mesh is made of three separate blocks in each of the vessel arms. The blocks join where the horizontal and vertical symmetry lines cross. The mesh quality is in general reasonable with mainly orthogonal cells that have low skewness, except in necks where the small vessel joins the main one and in the very central area where the mesh blocks join (cf. Fig. 9.11). Unlike in the case of the O-mesh discussed in Sec. 9.1.3, refinement does not make the skewness worse.

Being a structured mesh it is nearly everywhere perfectly regular with quadrilateral cells (when using a cell-centred discretisation) or quadrilateral blocks of cells around any given node (in case of using node-centred discretisations, cf. Secs. 7.1.2 and 7.4). The only irregularity in the grid is in the very centre where the blocks join. The vertex at the very centre is connected to six cells rather than four.

9.2.2 Flow physics and boundary conditions

Very differently from the inviscid aerofoil flow of the case in Sec. 9.1, this flow will exhibit very viscous flow with limited gradients and very gradual variations. Combined with the fact that there is no far-field in the domain that requires some of the resources, this flow can be adequately resolved with a moderate number of cells.

The primary feature that is to be investigated is whether or not there will be a flow separation at the upstream end of the neck. Fig. 9.12 sketches

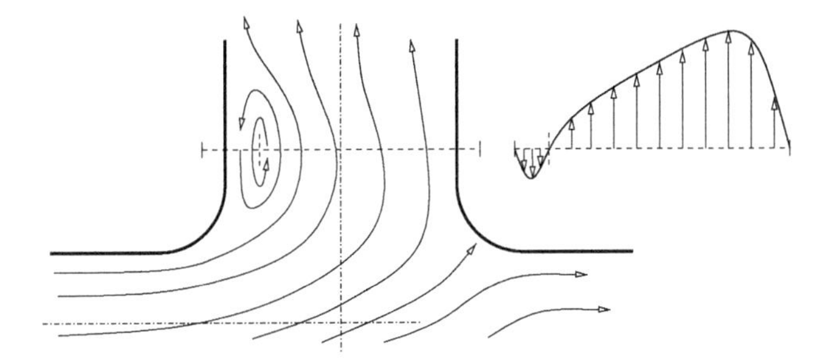

FIGURE 9.12
Sketch of a possible streamline pattern in the neck area with flow separation and recirculation bubble. The corresponding profile of y-velocity across the neck is drawn on the right.

the streamline pattern if a flow separation exists: the flow approaching the neck from the left is too fast to make the sharp turn. The inertia of the fluid particles makes the streamline separate away from the wall. A recirculation zone forms where fluid close to the wall slowly spins, with the circular motion in equilibrium between the acceleration by the faster inside fluid and the friction with the stationary wall.

Note that in two-dimensional flow the fluid in the recirculation bubble is completely separated from the other fluid: there is a limiting streamline that separates from the wall and reattaches further downstream. The particles in the recirculation zone are "trapped" as they can neither cross the wall, nor this limit streamline. This situation is an artefact of two-dimensional flow. In three dimensions the motion inside the separation bubble would be helical rather than circular: particles would be sucked into the recirculation zone e.g., at the symmetry line, "cork-screw" along and be spat out at the tips of the zone. The actual flow separation can possibly be quite different in an actual three-dimensional blood vessel.

The imposition of boundary conditions needs to be designed to control the split of mass flow between the two outlets as already discussed in Sec. 5.1.2. One option is to impose the incoming velocity at the inlet, outgoing velocity at the large outlet and the pressure at the small outlet. This choice allows the velocity to adjust at the small outlet to satisfy continuity, and allows pressure to adjust at the inlet and the large outlet to satisfy the momentum balance.

Imposing a uniform pressure of 0 Pa at the small outlet can be justified if the flow is fully developed, i.e., the final velocity profile in the cross section is established, all particles move parallel to the axis and the radial pressure gradient is zero. We will need to inspect the solution whether the flow is indeed fully developed.

At the inlet we impose a uniform velocity which is not correct: the flow arriving at that inlet cross section will have travelled through a significant length of vessel, and viscous effects and the no-slip condition will have produced a velocity profile with a peak at the centre. In this simple, straight geometry we know that the velocity profile should be parabolic and we could impose that profile, rather than uniform flow. However, a realistic blood vessel with a twisted centreline and irregular cross sections would not exhibit a known profile. In the general case we may impose a parabolic profile as a best guess, but we need to make the inlet length long enough for any errors about the profile of imposed quantities to have decayed before zones of interest are reached.

In this case we impose uniform flow. The no-slip condition at the wall and viscous effects will then slow the flow down near the wall and continuity in turn requires that the flow accelerates in the centre section. The velocity profile needs to change which is brought about by significant pressure gradients in the axial and radial direction. Once the flow is fully established, the profile no longer changes corresponding to zero radial pressure gradient, i.e., regularly spaced contours of pressure that are perpendicular to the vessel axis. We will need to inspect the solution whether the flow is indeed fully developed by the time it arrives at the neck.

We impose the velocity at the large outlet which determines the required mass flow split between the outlets. Again, we do not know what the velocity profile should be and for simplicity impose a uniform velocity. This, in fact, means that the flow needs to "un-develop", reducing velocity near the centreline and increasing velocity near the walls, which requires the appropriate radial and axial pressure gradients. Since we are not interested in the flow profile at the large outlet, those errors could be tolerated.

At the small outlet we impose a constant gauge pressure of zero,[5] which would only be correct if the flow is fully developed when it reaches the outlet and is unlikely to be satisfied here. We need to check whether the perturbation we introduce by the choice of both outlet conditions does not affect the velocity profile in the neck that is of most interest here.

9.2.3 Velocity and pressure fields

The computations are performed with OpenFOAM. Pressure fields for mesh levels 2-4 and for first- and second-order accurate discretisations are shown in Fig. 9.13. The artefacts of the uniform velocity conditions at the inlet on the left and the large outlet on the right are clearly visible. The flow in the main vessel downstream of the bifurcation appears to have a short section of straight pressure contours perpendicular to the vessel axis, indicating zero radial (cross-axis) pressure gradient and hence developed flow. This suggests that the errors of the velocity condition at the outlet should not affect the flow in the bifurcation zone.

[5]Working with gauge rather than atmospheric pressure minimises round-off error (cf. Sec. 4.2.1).

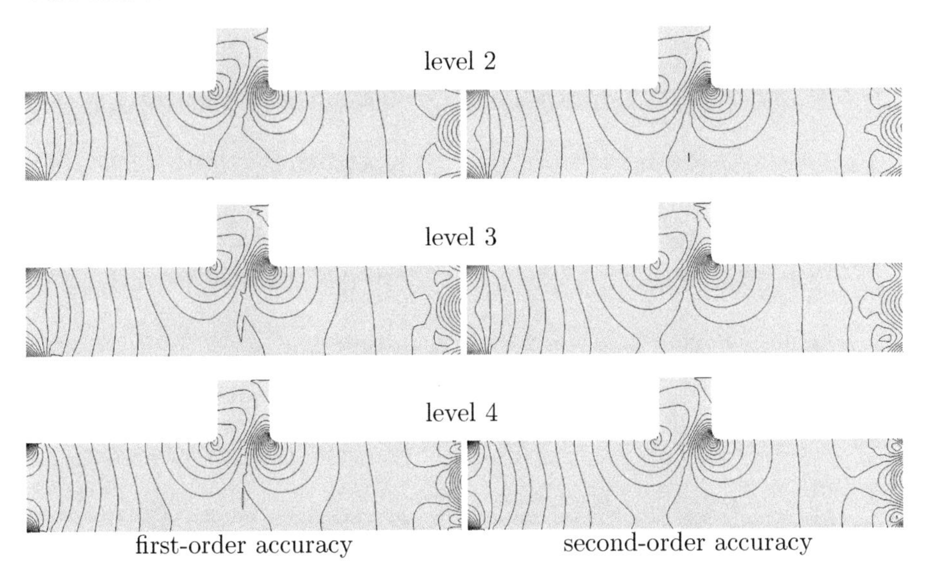

FIGURE 9.13
Pressure contours (50 levels from −0.02 Pa to 0.1 Pa) for bifurcation flow. Mesh levels 2, 3 and 4 from top to bottom. Left: first-order, right: second-order accuracy.

The same is not the case for the inlet. At no distance upstream of the bifurcation can we observe straight pressure contours normal to the axis. The flow has still not reached a fully developed profile by the time it reaches the bifurcation. Possibly the simulated length of inlet vessel should be extended to ensure that the flow profile just upstream of the bifurcation is not affected by the inlet profile.

The pressure condition at the small outlet forces a zero pressure gradient normal to the vertical symmetry axis; however, the rapid change in pressure field just upstream of the small outlet indicates that constant pressure is not a good approximation to the correct flow.

Other than in the regions close to where inlet/outlet boundary conditions are applied, the pressure field does not change significantly under mesh refinement or order increase. The pressure contours in the neck do shift around to some extent. Profiles of velocity will have to be investigated to obtain more detailed information.

In the first-order accurate solutions and the coarsest second-order solution a perturbation in the pressure contours can be observed at the very centre. This is due to the irregularity of the mesh.

Velocity magnitude contours are shown in Fig. 9.14. The development of the flow downstream of the inlet and the "un-development" just upstream of the large outlet can clearly be seen. The flow in the neck is biased to the right-hand side of the small vessel; at the left a zone of very low velocity can

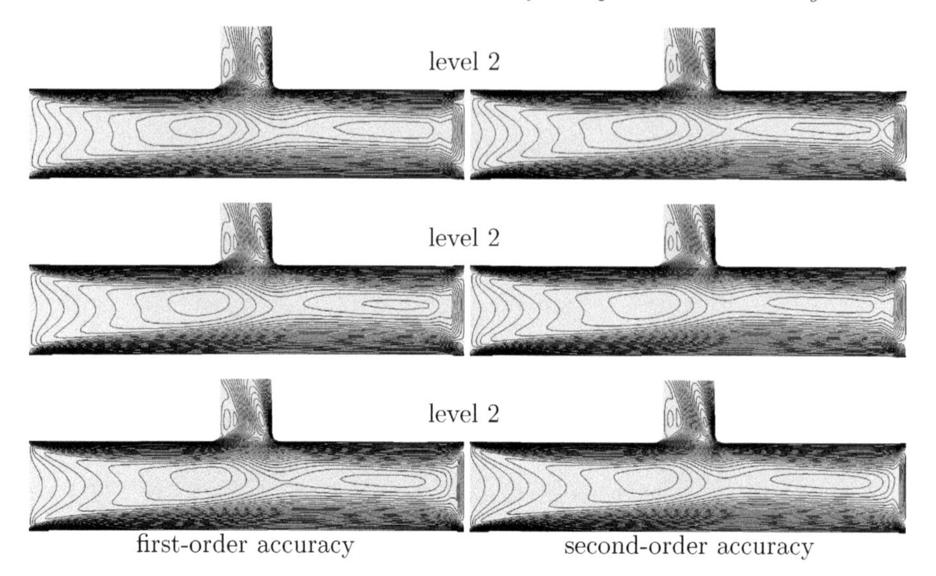

FIGURE 9.14
Velocity magnitude contours (50 levels from 0 m/s to 0.5 m/s) for bifurcation flow. Mesh levels 2, 3 and 4 from top to bottom. Left: first-order, right: second-order accuracy.

be seen which may be a flow separation, but could also be a zone of very low upward velocity. Contour or profile plots of y-velocity are needed to confirm this.

9.2.4 Velocity profile in the neck

Velocity profiles on a line just above the neck are shown in Fig. 9.15 for all levels and first- and second-order accuracy. All simulations exhibit a distinct recirculation near the left wall, identified by the negative y-velocity there, with the second-order simulations showing a slightly slower recirculation compared to the first-order ones in Fig. 9.16 (a). Peak velocity in Fig. 9.16 (b) is increasing with mesh refinement and increased order of accuracy. However, the differences between the second-order accurate solutions on level 3 and level 4 are very small and level 3 could be considered mesh converged.

9.2.5 Effect of outlet boundary condition

To assess what effect the boundary condition at the small outlet has on the velocity profile at the neck, we switch the pressure and velocity boundary conditions between the large and the small outlet. The velocity contours in Fig. 9.17 show how the fixed velocity condition distorts the velocity field near the outlets where that condition is imposed. The pressure outlet condition

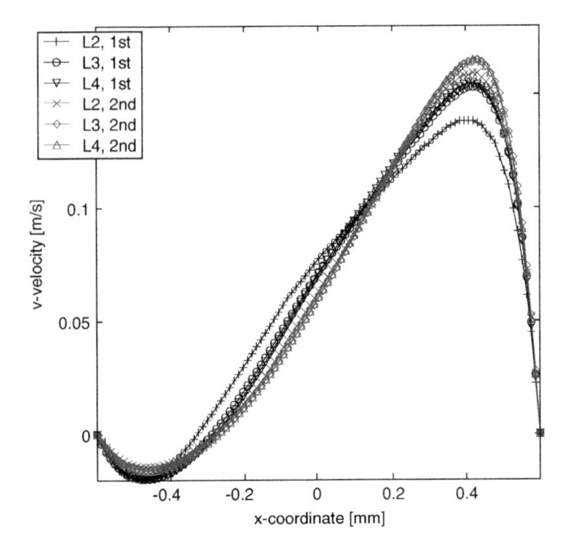

FIGURE 9.15
Profile of y-velocity just above the neck.

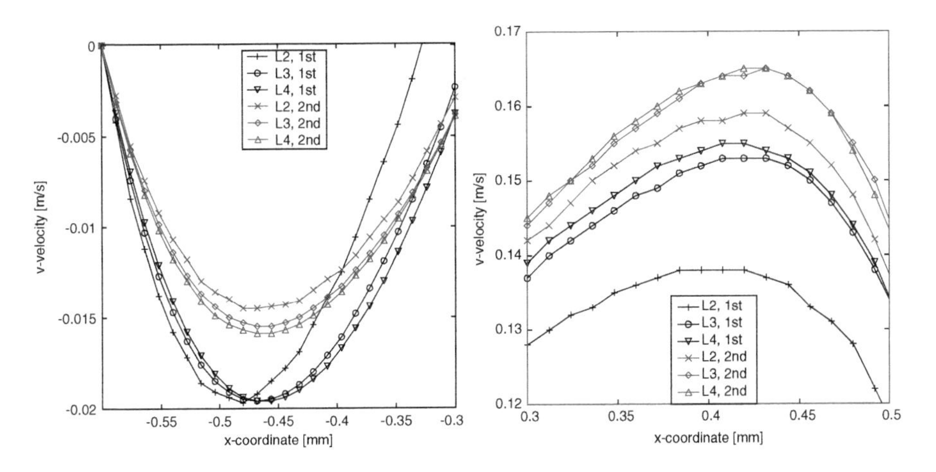

FIGURE 9.16
Profile of y-velocity just above the neck, detail (a) in the recirculation zone near the left wall and (b) the velocity peak near the right wall.

is still not accurate, but has a smaller effect on the flowfield near the outlet compared to the velocity condition.

Given the substantial change in flow at the small outlet, the velocity and pressure contours in the neck show remarkably little change. Profiles of y-velocity are given in Fig. 9.18 and show very similar characteristics: both

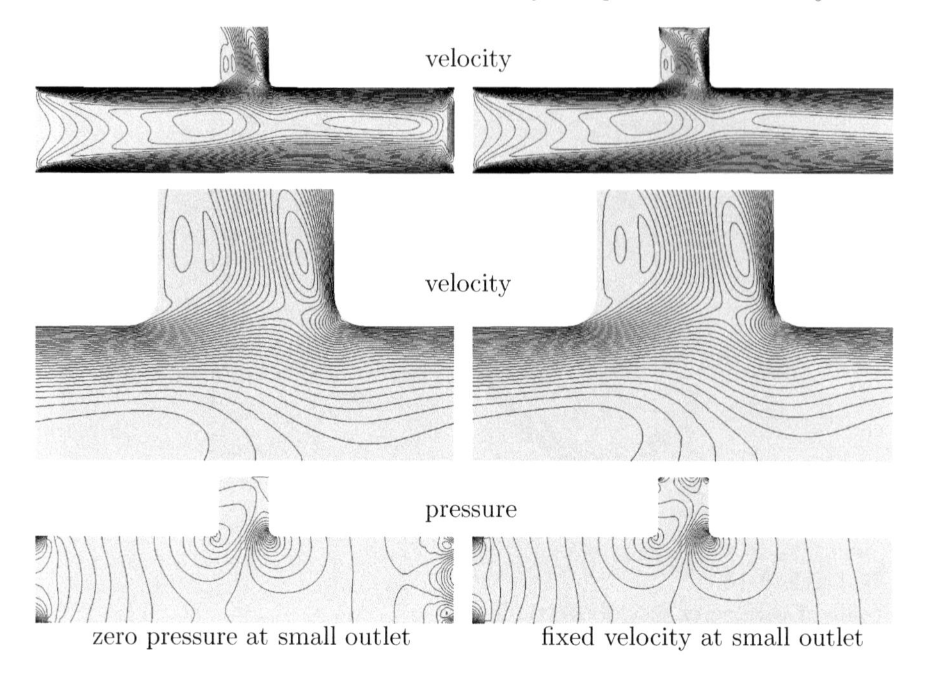

FIGURE 9.17
Comparison of flowfields with pressure outlet condition at the small outlet (left) and at the large outlet (right); level 4 mesh with second-order accurate discretisation. Top and middle: velocity magnitude contours (50 levels from 0 m/s to 0.5 m/s), bottom: pressure contours (50 levels from −0.2 Pa to 0.1 Pa).

profiles exhibit a very similar separation zone, but there are slight variations in the minimal and maximal velocity values.

It depends on the required accuracy whether the deviations that are observed are acceptable or whether the length of the small vessel needs to be extended to reduce the influence of any boundary condition errors.

9.3 Aerofoil in 2-D, viscous flow

In Sec. 9.1 we computed the flow over the well-known NACA0012 aerofoil in inviscid flow. To be able to compare to wind tunnel experiments let us refine the modelling for that aerofoil to include the effects of viscous, turbulent flow. There is a wealth of experimental data, as well a study of the performance of a range of turbulence models on this case by NASA [10].

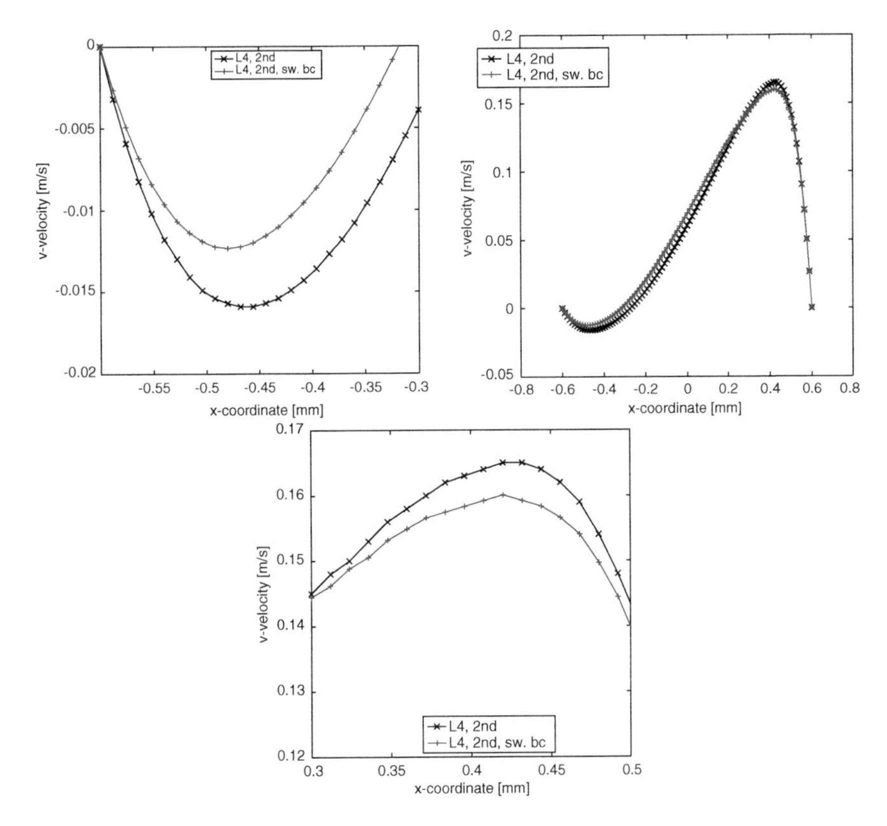

FIGURE 9.18
Profile of y-velocity just above the neck (middle), Left detail: in the recirculation zone near the left wall. Right: the velocity peak near the right wall.

9.3.1 Flow physics

The experiments were run at a Mach number of 0.15, and hence can be considered incompressible. The Reynolds number is between $3{\cdot}10^6$ and $6{\cdot}10^6$. To match the Reynolds number $\mathrm{Re} = ul/\nu$ we can adjust either or all of the freestream velocity u, the chord length l or the viscosity ν. Here we choose to leave viscosity unaltered at $\nu = 15{\cdot}10^{-6}\,\frac{\mathrm{m}^2}{\mathrm{s}}$ and use the same inflow velocity as for the inviscid case in Sec. 9.1 at $u_\infty = 59\,\frac{\mathrm{m}}{\mathrm{s}}$. To match a Reynolds number of $\mathrm{Re} = 6 \cdot 10^6$ we then need to adjust the chord length of the mesh to be $l = 1.52\,\mathrm{m}$.

We keep the outer boundary at 20 chord lengths, same as for the inviscid case of Sec. 9.1. A larger distance would be preferable, but with 20 chord lengths the error in lift due to this will be small.

The tangential mesh spacing is governed by the tangential pressure and velocity changes which are very similar to the ones of the inviscid aerofoil case of Sec. 9.1. To limit the computational effort, we shall limit ourselves to 256 cells in the profile direction, knowing all well that this mesh resolution has shown to be not grid-converged in the inviscid studies. Lift is of interest here. We, hence, use the C-mesh topology which avoids poor mesh quality at the trailing edge (cf. Sec. 9.1.5).

The flow will exhibit turbulent boundary layers and a wake. The mesh spacing normal to the wall needs to be adjusted as discussed in 6.7. The NASA website [10] shows how the results differ whether transition from laminar to turbulent is allowed to occur naturally, or whether the flow is "tripped" to be turbulent from the beginning. As prediction of transition is very difficult, we shall assume that the flow is fully turbulent from the stagnation point.

For low angles of attack we can expect the flow to remain attached: both streamlines from the stagnation point will remain attached to the profile on suction and pressure side to leave the profile at the trailing edge. With an increase in angle of attack, lift will increase at a near constant rate. For a flat plate in inviscid flow, analytic solutions give this rate as $\frac{\partial c_L}{\partial \alpha} = 2\pi\alpha$ with the angle of attack α measured in radian.

The drag of the aerofoil in this case consists of two main components. At zero angle of attack drag is non-zero, corresponding to the friction over the wetted surface, hence called "skin friction drag". With increasing angle of attack, drag increases at a rate approximately proportionally to α^2. This drag increase is due to the loss of momentum in the flow which affects the pressure field. It is also called "pressure drag".

At higher angles of attack the flow on the upper, suction surface will separate. The experiments identify this as an abrupt drop in lift and increase in drag at an angle of 15-17° for this profile and Reynolds number. At these high angles of attack the flow over the suction surface will have seen a significant acceleration and pressure drop to a strong suction peak. Downstream of that the flow experiences an *adverse pressure gradient* that decelerates the flow. If the flow has insufficient inertia to maintain the trajectory along the aerofoil surface, streamlines are deflected away from the zone of increasing pressure, hence away from the body.

To satisfy continuity, the flow downstream of the separating streamline needs to change dramatically, since there is no longer incoming flow from upstream. This forces the creation of a recirculation bubble where in the two-dimensional case streamlines form closed loops, or in the three-dimensional case, streamlines may form helical eddies with flow e.g., entering the helix at one end and exiting at the other.

This change does not happen gradually, but occurs as a sudden change. In aeronautics the event is called *stall*. It is the cause of many aircraft accidents. With standard techniques for turbulence modelling it is very difficult to predict stall.

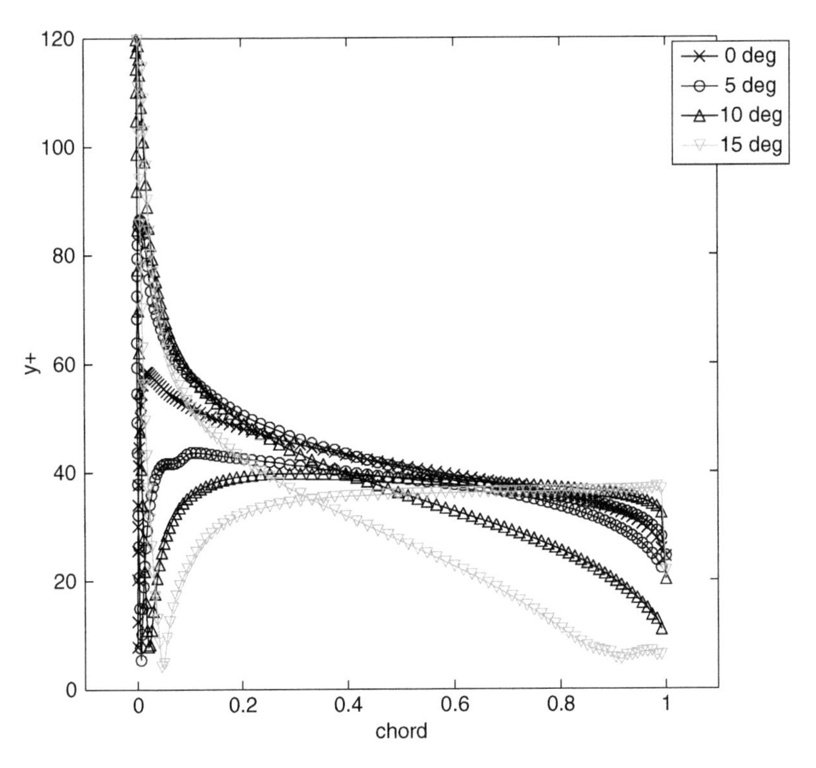

FIGURE 9.19
y+ values for $k-\varepsilon$ and Spalart-Allmaras turbulence models for a range of angles of attack.

9.3.2 Turbulence modelling

We shall use standard RANS turbulence modelling with the k–ε and Spalart-Allmaras turbulence models. To keep computational cost reasonable we choose the wall-function approach (cf. Sec. 6.7). The graphs include comparison data with simulations from the NASA website [10] that use a low-Reynolds approach with resolution of the sublayer.

We can expect the RANS models with wall functions to work reasonably well for the lower angles of attack. The assumptions of fully developed flow with small pressure gradients will not be not valid in regions of flow separation, and as the results will show, stall is not captured by the RANS models with any level of accuracy.

Fig. 9.19 shows the distribution of y^+ for a range of angles of attack. The physical spacing y of the first cell of the wall has been kept constant around the profile to keep the mesh generation simple. This, in turn, implies that the y^+ will vary with the wall shear stress τ. An attempt was made to keep y^+ as much as possible in the range of 30-60. y^+ drops below 30 in the case of 10°

angle of attack from around 70% chord, and for the case of 15° from around 45% chord. The Spalart-Allmaras model is often implemented in a way that it allows to smoothly refine mesh from log-layer to sublayer. The Fluent solver used here also offers this. We are, hence, not concerned about the y^+ dropping below the log-layer. However, this would not be the case for the $k-\varepsilon$ model, where we would have needed to adjust the mesh to raise y^+ above 30.

9.3.3 Flow results

Fluent (Version 14) has been used for the computations, using a standard SIMPLE pressure correction discretisation for incompressible flows and second-order accurate schemes for pressure correction, momentum and turbulent transport equations. Boundary conditions for velocity and pressure are imposed as for the inviscid case of Sec. 9.1. Additionally we need to provide inflow values for the turbulence variable. Following Fluent's guidance for boundary layer flows we specify the turbulence intensity and the turbulent length scale. A typical turbulent intensity is 5%. For boundary layer flows Fluent suggests to specify the lengthscale taken as $l = .4\delta$ where δ is the boundary layer thickness. Using a flat plate approximation for δ at 1,m of running length, we obtain a turbulent length scale of 0.00674 m.

Experimental values for pressure coefficient $c_P = \frac{p}{\frac{1}{2}\rho u_\infty^2}$ are published [10] for the angles of attack of 0°, 10° and 15°. Fig. 9.20 shows the velocity magnitude contours and c_P profiles for these angles. We can observe how the boundary layer and wake increase in thickness with increasing angle of attack. The stronger suction peak at the higher angles of attack implies a more severe adverse pressure gradient from the suction peak to the trailing edge. This, in turn, leads to a thicker boundary layer and ultimately to flow separation and stall at even higher angles of attack.

The experimental pressure measurements, which are available for the suction side only, are also shown in Fig. 9.20. For these angles of attack the agreement of numerical simulation and experiment is very good.

9.3.4 Lift and drag

The values of lift and drag for the simulations with the Spalart-Allmaras model are compared in Fig. 9.21 against the available experimental NASA data [10]. For comparison, computations with the k–ε model are included. For the k–ε-model incoming turbulence was specified through turbulent intensity at 5% and turbulent viscosity ratio at a value of 10.

Both the Spalart-Allmaras and the k–ε turbulence models show very good agreement for an angle of attack of up to about 14°. Neither model is able to predict the separation correctly to any degree of accuracy. The k–ε model with a freestream specification of turbulence using the turbulent viscosity ratio of 10 (labelled "turb. ratio" in Fig. 9.21 (a)) fails to predict any separation at all. The lift curve does level off at around 20°, but inspection of the flow field

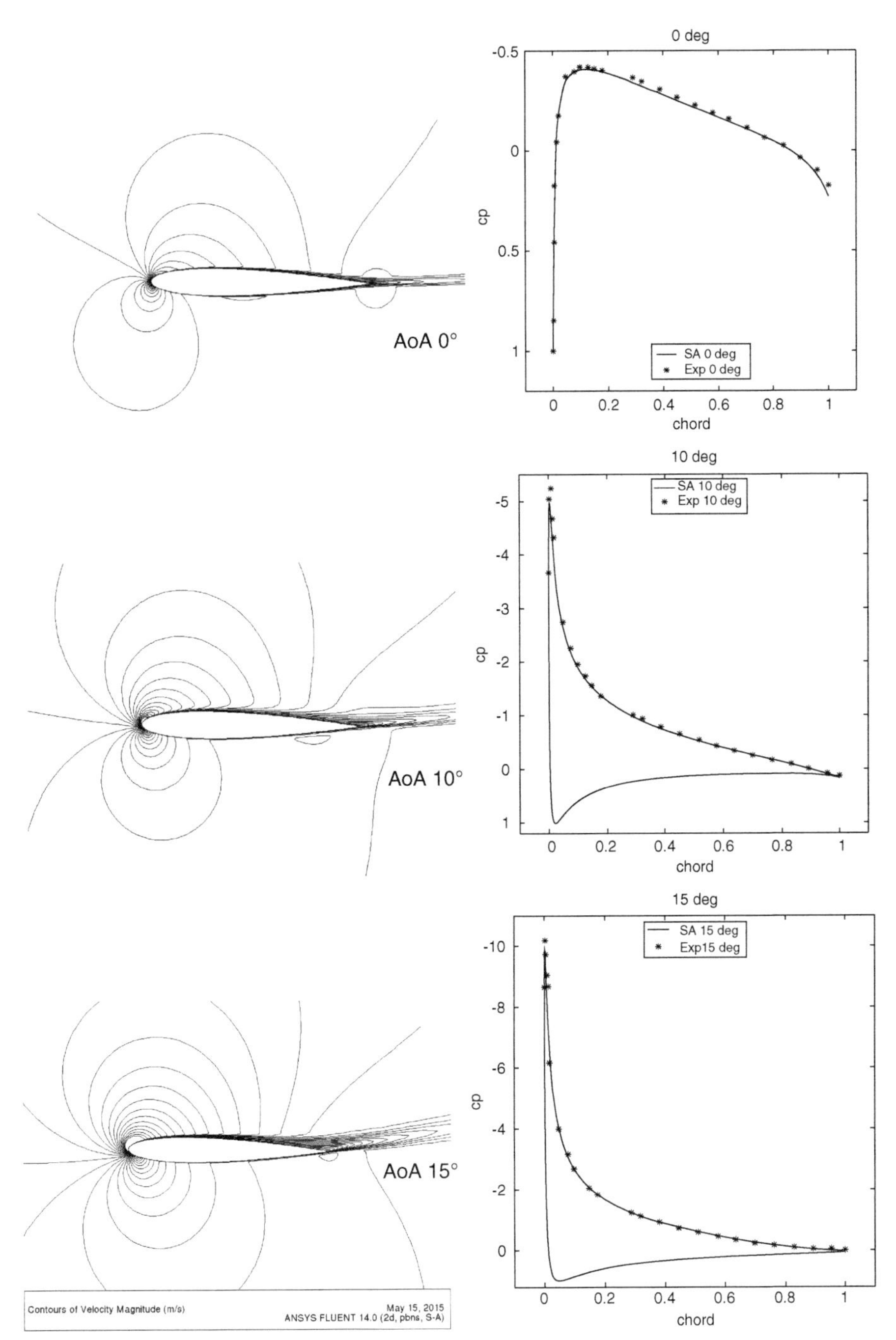

FIGURE 9.20
Contours of velocity magnitude (left) and pressure coefficient c_P of simulation and experimental data (right) for 0°, 10° and 15° angle of attack.

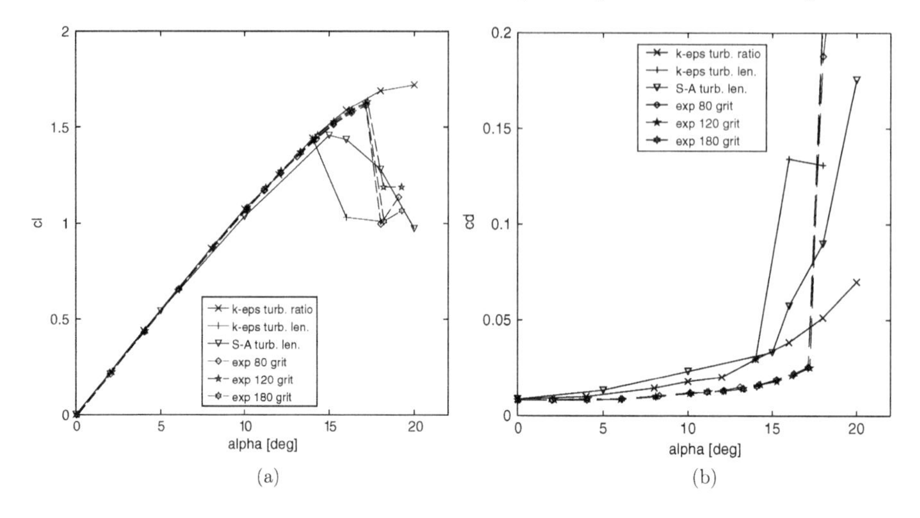

FIGURE 9.21
Lift (a) and drag (b) vs. angle of attack.

shows a rather large stagnant zone on the suction side, but no reverse flow. This inlet condition clearly is too viscous for this situation.

The Spalart-Allmaras model and the $k-\varepsilon$ model with a freestream specification using the turbulent lengthscale (labelled "turb. len." in Fig. 9.21) fare much better. Both predict a separation. However, in both cases the separation occurs at an angle of attack that is too low. The Spalart-Allmaras model also predicts a gradual decrease in lift, rather than the sudden drop as observed in stall.

The change of lift behaviour observed when changing the inlet condition for the freestream level of turbulence for the $k-\varepsilon$ is significant. It is very difficult to determine an objective value for the required quantities. By careful adjustment of the freestream turbulence we may be able to tune the model into producing stall for this case; however, the required inlet conditions will be very different for another case with separation. If RANS models are used to predict separated flows, these parameters have to be adjusted based on the experience of the user with their cases and must be carefully validated against experimental results. The range of validity of the model will be very limited to flows that are closely related to the ones the model has been calibrated for. It is to be noted, however, that the turbulence models are reasonably insensitive against parameter changes as long as the flow remains attached; unfortunately, though, many industrial flows do exhibit separation zones.

The drag at zero angle of attack is predicted reasonably well, as shown in Fig. 9.21 (b). The drag then increases too rapidly with angle of attack for all models, and more so in this case for the Spalart-Allmaras model. Similarly to the lift, neither model is able to predict the rapid increase of drag occurring at stall.

Better results can be achieved with more care, e.g., using a finer mesh and replacing the wall function approach with a resolution into the viscous sublayer with $y^+ \leq 1$. This will improve, in particular, the drag prediction. However, similar conclusions would be drawn as for the lift, and also apply to all other RANS turbulence models: RANS models are not able to robustly predict flow separation. If the flow that is simulated contains separation zones, the user must carefully examine whether or not that part of the flow is predicted with sufficient accuracy, and what effect the errors for that part of the flow have on overall flowfield and the objective functions such as lift and drag.

10

Appendix

10.1 Finite-volume implementation of 2-D advection

```
%% function circular_advection ()
%% circular advection in 2d using finite volumes.

%% Last update:
%% 24Aug14; derived from advect_fv, JDM

%% Tidy up from previous run.
clear;
close all;


%%%%%%%%%%%%%%%%%%%%%%%%%%%%%%%%%%%%%%%%%%%%

function u = initial_solution ( NC )
  %% initialise the scalar field u to zero

  %% input
  %% NC: number of cells in x,y

  %% output
  %% u: scalar field initialised to zero.

  %% declare and initialise the field u
  u = zeros(NC,NC) ;
end

%%%%%%%%%%%%%%%%%%%%%%%%%%%%%%%%%%%%%%%%%%%%

function res = residual ( u, NC ) ;
  %% calculate the residual, i.e. the rate of
  %% change or the flux divergence.

  %% input
  %% u: scalar field u
  %% NC: number of cells in x,y
```

```
%% output
%% res: flux divergence/residual of u


%% Initialise the residual to zero.
res = zeros(NC,NC) ;

h = 1/NC ; % mesh width

%% compute fluxes of internal horizontal faces.
for i=1:NC
  for j=1:NC-1
    %% cell south of the interface
    i1 = i;
    j1 = j;
    %% cell north of the interface:
    i2 = i ;
    j2 = j+1;
    %% normal direction (outward for the southern cell)
    n = [0,1] ;

    %% coordinates of the upwind/southern cell
    xi = h/2 + (i1-1)/NC ;
    yi = h/2 + (j1-1)/NC  ;
    %% advection speed at the upwind/southern cell
    a = [ yi, 1-xi ] ;

    %% flux
    f = a*u(i1,j1) ;
    %% flux projected onto the face
    fns = f*n'*h ;

    %% add to the left, subtract from the right for a>0.
    res(i1,j1) = res(i1,j1) + fns ;
    res(i2,j2) = res(i2,j2) - fns ;
  end
end


%% internal vertical faces
for i=1:NC-1
  for j=1:NC
    %% cell west of the interface
    i1 = i;
    j1 = j;
    %% cell east of the interface:
    i2 = i+1 ;
    j2 = j;
    %% normal direction (outward for the western cell)
```

```
    n = [1,0] ;


    %% coordinates of the upwind/southern cell
    xi = h/2 + (i1-1)/NC ;
    yi = h/2 + (j1-1)/NC ;
    %% advection speed at the upwind/southern cell
    a = [ yi, 1-xi ] ;

    %% flux
    f = a*u(i1,j1) ;
    %% flux projected onto the face
    fns = f*n'*h ;

    %% add to the left, subtract from the right for a>0.
    res(i1,j1) = res(i1,j1) + fns ;
    res(i2,j2) = res(i2,j2) - fns ;
  end
end



%% boundary condition left, x=0:
i1 = 1 ;
for j1=1:NC
  %% boundary condition
  u_bc = 0 ;
  %% normal direction (outward for the eastern cell)
  n = [-1,0] ;

  %% coordinates of the boundary face
  xi = 0 ;
  yi = h/2 + (j1-1)/NC ;
  %% advection speed at the boundary face
  a = [ yi, 1-xi ] ;
  %% flux
  f = a*u_bc ;
  %% flux projected onto the face
  fns = f*n'*h ;

  %% this adds zero, just shown for completeness
  res(i1,j1)    = res(i1,j1) + fns ;
endfor

%% boundary condition bottom
j1 = 1 ;
for i1=1:NC
  %% boundary condition
  if i1 < NC/4
```

```
    %% left quarter, 0<x<0.25, y=0:
    u_bc = 0 ;
  elseif i1 > 3*NC/4
    %% right quarter, 0.75<x<1, y=0:
    u_bc = 0 ;
  else
    %% centre half, 0.25<x<0.75, y=0:
    u_bc = 1 ;
  endif
  %% normal direction (outward for the northern cell)
  n = [0,-1] ;

  %% coordinates of the boundary face
  xi = h/2 + (i1-1)/NC ;
  yi = 0 ;
  %% advection speed at the boundary face
  a = [ yi, 1-xi ] ;

  %% flux
  f = a*u_bc ;
  %% flux projected onto the face
  fns = f*n'*h ;

  res(i1,j1)   = res(i1,j1) + fns ;
endfor


%% boundary condition right, x=1:
i1 = NC ;
for j1=1:NC
  %% boundary condition: outflow, use internal value.
  u_bc = u(i1,j1) ;
  %% normal direction (outward for the western cell)
  n = [1,0] ;

  %% coordinates of the boundary face
  xi = h/2 + (i1-1)/NC ;
  yi = h/2 + (j1-1)/NC ;
  %% advection speed at the boundary face
  a = [ yi, 1-xi ] ;

  %% flux
  f = a*u_bc ;
  %% flux projected onto the face
  fns = f*n'*h ;

  res(i1,j1)   = res(i1,j1) + fns ;
endfor
```

```
  %% boundary condition top, y=1:
  j1 = NC ;
  for i1=1:NC
    %% boundary condition: outflow, use internal value.
    u_bc = u(i1,j1) ;
    %% normal direction (outward for the southern cell)
    n = [0,1] ;

    %% coordinates of the boundary face
    xi = h/2 + (i1-1)/NC ;
    yi = h/2 + (j1-1)/NC ;
    %% advection speed at the boundary face
    a = [ yi, 1-xi ] ;

    %% flux
    f = a*u(i1,j1) ;
    %% flux projected onto the face
    fns = f*n'*h ;

    res(i1,j1)   = res(i1,j1) + fns ;
  endfor

end

%%%%%%%%%%%%%%%%%%%%%%%%%%%%%%%%%%%%%%%%%%%%%%

function [u, dt, rms, u_min, u_max] =
  explicit_timestep ( u, res, cfl, a_max, NC )
  %% explicit update using forward discretisation
  %% in time (forward Euler).

  %% input
  %% u: scalar field at time t
  %% res: residuals (rate of change, flux divergence)
  %%      in each control volume
  %% cfl: cfl number to scale the maximally allowable
  %%      timestep
  %% a_max: maximal advection speed in the domain
  %% NC: number of cells in x,y

  %% output
  %% u: updated scalar field at time t+dt
  %% dt: timestep
  %% rms: root mean square of the residual
  %% u_min: minimal value of u at t+dt
  %% u_max: maximal value of u at t+dt


  %% timestep is limited by fastest advection speed u_max
```

```
  h = 1/NC ; % mesh width
  A = h^2 ;  % cell area
  dt = cfl*h/a_max ;% maximally allowable time-step

  %% Time discretisation/Update
  % track the rms of the residuals to measure convergence
  rms = 0. ;
  for i=1:NC
    for j=1:NC
      %% Explicit time-stepping
      u(i,j)=u(i,j)-dt/A*res(i,j) ;

      %% monitor convergence, use the root mean
      %% square of the residual
      rms = rms + res(i,j)^2 ;
    endfor
  endfor

  ## root mean square of the residual.
  rms = sqrt(rms/NC^2) ;
  u_min = min(min(u)) ;
  u_max = max(max(u)) ;

end

%%%%%%%%%%%%%%%%%%%%%%%%%%%%%%%%%%%%%%%%%%%%%%%%%%

function plot_field ( u, NC, mode )
  %% plot contours of a scalar field

  %% input
  %% u: scalar field
  %% NC: number of cells in x,y
  %% mode: 0: contours,
  %%       1: carpet plot,
  %%       3: outlet profile vs. exact solution.

  %% field of grid node coordinates (not cell centres.)
  x = zeros( NC+1, NC+1 ) ;
  y = zeros( NC+1, NC+1 ) ;


  %% fill coordinate and temperature fields. The grid node
  %% numbered i,j is to the bottom left (south west) of
  %% the cell centre numbered i,j.
  for i = 1:NC+1
    for j = 1:NC+1
      x(i,j) = (i-1)/NC ;
      y(i,j) = (j-1)/NC ;
```

```
  end
end
uu = zeros(NC+1,NC+1) ;

%% interpolate cell centre values to grid node values
%% internal grid nodes are averaged from the four surrounding
%% nodes, i.e., each cell centre contributes with 1/4 to each
%% surrounding grid node. Correct for boundaries.
for i = 1:NC
  for j = 1:NC
    uu(i,j)     = uu(i,j)     + u(i,j) ;
    uu(i+1,j)   = uu(i+1,j)   + u(i,j) ;
    uu(i,j+1)   = uu(i,j+1)   + u(i,j) ;
    uu(i+1,j+1) = uu(i+1,j+1) + u(i,j) ;
   end
 end

%% internal grid nodes received 4 contributions.
for i = 2:NC
  for j = 2:NC
    uu(i,j) = uu(i,j)/4 ;
  end
end

%% edge grid nodes received 2 contributions
for i = 2:NC
  uu(i,    1) = uu(i,   1)/2 ; % bottom
  uu(i, NC+1) = uu(i,NC+1)/2 ; % top
end
for j = 2:NC
  uu(1,   j) = uu(1,  j)/2   ; % left
  uu(NC+1, j) = uu(NC+1,j)/2 ; % right
end

if mode == 2
 %% mesh plot
 mesh ( x(:,1), y(1,:), uu ) ;

elseif mode == 3
  %% target profile
  yt = [ 0, .25, .25,  .75, .75, 1. ] ;
  ut = [ 0.  0., 1.0, 1.0, 0., 0. ] ;

  plot ( y(NC,:), uu(NC,:), '-x',  yt, ut, '-*' ) ;
  axis ( [-.2, 1.2, -.2, 1.2], "square" ) ;
  legend ( 'u|x=1', 'exact|x=1' )
  title ( 'Profiles of simulation and exact solution at x=1' )

else
```

```
    %% contour plot
    lvl = linspace(0,1,11) ;
    contourf  ( x, y, uu, lvl ) ;
    axis equal ;

    title ( 'Passive scalar u at grid nodes' ) ;
  endif

end




%%%%%%%%%%%%%%%%%%%%%%%%%%%%%%%%%%%%%%%%%%%%%%%%%%%%
%% main
%%%%%%%%%%%%%%%%%%%%%%%%%%%%%%%%%%%%%%%%%%%%%%%%%%%%


% How many cells in each coordinate direction should be used?
NC=input(' Number of cells NC in each dir = ?');
if isempty ( NC ) NC = 4 ; end

% How many timesteps
nt=input(' Number of timesteps nt= ?');
if isempty ( nt ) nt = 1 ; end


% Choose a CFL number
cfl=input(' CFL number cfl= ') ;
% maximum advection speed in the top left corner limits the timestep
if isempty ( cfl ) cfl = 0.5 ; end
%% maximal advection speed is found in the top left corner,
%% this limits the time-step in the explicit time-stepping.
a_max = sqrt(2) ;



# intialise the field.
u = initial_solution ( NC ) ;


%% Index n loops over all timesteps nt
t_total = 0.0 % time covered by the simulation.
nplot = -1 ; % number of plots of the field so far.
for n=1:nt

  %% Space discretisation: loop over all faces and
  %% accumulate the flux balance, or residual.
```

```
  res = residual ( u, NC ) ;

  [u, dt, rms, u_min, u_max] =
    explicit_timestep ( u, res, cfl, a_max, NC ) ;
  t_total = t_total + dt ;
  fprintf ( ' iter %3d, time %6.3f, log of RMS:'...
            ' %7.3f, u_min %6.3f, u_max %6.3f.\n',
            n, t_total, log10(rms), u_min, u_max ) ;

  %% The magnitude of the advection speed is r=x^2+y^2, so
  %% a quarter revolution should take pi/2 in time.
  %% Plot the field every 10/th of pi/2.
  mplot = 5 ;
  if ( floor( mplot*t_total/pi*2 ) > nplot )
    nplot = nplot+1 ;
%     hold off
    plot_field ( u, NC, 1 ) ;
%     hold on ;
    title ( sprintf ( 'solution at time t=%6.3f', t_total ) ) ;
    pause ( 0.02 ) ;
  end
end


plot_field ( u, NC, 1 ) ;
title ( sprintf ( 'circular advection, %dx%d cells',NC,NC ) ) ;
pause ;
%% matlab syntax for jpg is -djpeg
print ( '-dpng','circular_advection_field.png') ;


%% mesh plot, not very useful in octave
%%plot_field ( u, NC, 2 ) ;
%% pause ;

%% plot exit profile, exact solution
plot_field ( u, NC, 3 ) ;
print ( '-dpng','circular_advection_prof.png') ;
pause ;
```

Bibliography

[1] Richard H. Pletcher, John C. Tannehill, and Dale Anderson. *Computational Fluid Mechanics and Heat Transfer*. CRC Press, 3rd ed. edition, 2012.

[2] C. Hirsch. *Numerical Computation of Internal and External Flows, Vol. I and II.* Wiley, 1990.

[3] J.H. Ferziger and M. Perić. *Computational Methods for Fluid Dynamics.* Springer, "New York, NY, USA", 2002.

[4] J. Blazek. *Computational Fluid Dynamics: Principles and Applications.* Elsevier, Amsterdam, 2nd. ed. edition, 2005.

[5] Randall J. LeVeque. *Finite Volume Methods for Hyperbolic Problems.* Cambridge University Press, Cambridge, UK, 2004.

[6] F.M. White. *Viscous Fluid Flow.* McGraw-Hill, 1974.

[7] I. Babuška and A.K. Aziz. On the angle condition in the Finite-Element method. *SIAM J. Num. Anal.*, 13(2):214–26, 1976.

[8] J. F. Thompson and N. P. Weatherill. Aspects of numerical grid generation: Current science and art. *AIAA invited paper*, August 9-11 1993. 11th AIAA Applied Aerodynamics Conference.

[9] J. F. Thompson, B. K. Soni, and N. P. Weatherill. *Handbook of Grid Generation.* Boca Raton, Fla., CRC Press, 1999.

[10] Langley Research Centre. Turbulence modeling resource. http://turbmodels.larc.nasa.gov/index.html.

Index

T - #0230 - 160425 - C0 - 234/156/11 - PB - 9781482227307 - Gloss Lamination